THE ELECTRONIC STRUCTURE OF
POINT DEFECTS

SERIES DEFECTS IN CRYSTALLINE SOLIDS

Editors:

S. AMELINCKX
R. GEVERS
J. NIHOUL

Studiecentrum voor kernenergie, Mol,
and
University of Antwerpen, Belgium

Vol. 1

R. S. NELSON: THE OBSERVATION OF ATOMIC COLLISIONS
IN CRYSTALLINE SOLIDS

Vol. 2

K. M. BOWKETT AND D. A. SMITH: FIELD-ION MICROSCOPY

Vol. 3

H. KIMURA AND R. MADDIN: QUENCH HARDENING IN METALS

Vol. 4

G. K. WERTHEIM; A. HAUSMANN AND W. SANDER:
THE ELECTRONIC STRUCTURE OF POINT DEFECTS

G. K. Wertheim: The study of crystalline defects by Mössbauer effect

A. Hausmann and W. Sander: Study of structural defects by spin resonance methods

NORTH-HOLLAND PUBLISHING COMPANY - AMSTERDAM · LONDON

THE ELECTRONIC STRUCTURE
OF POINT DEFECTS

AS DETERMINED BY MÖSSBAUER
SPECTROSCOPY AND BY
SPIN RESONANCE

G. K. WERTHEIM

Bell Telephone Laboratories
Murray Hill, N.J.
U.S.A.

A. HAUSMANN
and
W. SANDER

2. Physikalisches Institut der
Rheinisch-Westfälischen Technischen Hochschule
51-Aachen
Germany.

1971

NORTH-HOLLAND PUBLISHING COMPANY - AMSTERDAM · LONDON
AMERICAN ELSEVIER PUBLISHING COMPANY, INC. - NEW YORK

Library of Congress Catalog Card Number: 74-146194
North-Holland ISBN for the series: 0 7204 7050 3
North-Holland ISBN for this volume: 0 7204 1754 6
American Elsevier ISBN: 0 444 10125 x

Publishers:

North-Holland Publishing Company – Amsterdam
North-Holland Publishing Company Ltd. – London

Sole distributors for the U.S.A. and Canada:

American Elsevier Publishing Company, Inc.
52 Vanderbilt Avenue
New York, N.Y. 10017

Printed in The Netherlands

PREFACE

The Mössbauer effect has been originally studied by nuclear physicists but it rapidly gained importance in the field of solid state physics and it is fair to say that nowadays the main applications of the Mössbauer effect are in this area. Crystal defects resulting from atomic displacements by irradiation with energetic particles or by plastic deformation have relatively little been investigated by this method until now. On the other hand, ionization effects as well as the presence of chemical impurities and their interactions with physical defects are increasingly being studied by means of Mössbauer spectroscopy.

The most detailed results concerning the electronic structure of point defects in insulators have undoubtedly been obtained by the use of spin resonance techniques. Especially since the development of the ENDOR technique it became possible to observe in great detail the hyperfine structure of the spectra and hence one could determine the wave functions of unpaired electrons or holes. The detailed structure of a number of colour centres in alkali halides and in other insulators has been determined in this way. The method is clearly limited to centres that contain unpaired electrons or holes. Although the application of nuclear magnetic resonance (NMR) to the study of defects is severely limited on account of the much lower sensitivity of this method with respect to ESR, a brief outline of its specific applications in this field seemed to be entirely justified.

The numerous examples of applications treated in both parts of this volume clearly illustrate the important contribution of the Mössbauer and spin resonance methods to a better knowledge of the nature and the behaviour of defects in crystalline solids. We hope that this volume may stimulate effective research by these methods which form an invaluable complement to the more conventional ones used in this field.

We take this opportunity of thanking the authors of both monographs for the excellent way in which they accomplished the difficult task of covering a rather young branch in the field of structural defects in solids. We wish to extend our thanks to North-Holland Publishing Company, and especially to Drs. W. H. Wimmers, for their initiative to start with the series.

S. Amelinckx
R. Gevers
J. Nihoul

CONTENTS

PART I

THE STUDY OF CRYSTALLINE DEFECTS BY MÖSSBAUER EFFECT

G.K. WERTHEIM

Bell Telephone Laboratories
Murray Hill, N.J.
U.S.A.

1 | INTRODUCTION TO THE MÖSSBAUER EFFECT

Prior to the work of Rudolf Mössbauer in 1958 one did not think of the gamma ray part of the electromagnetic spectrum as a region where high resolution spectroscopy was feasible. There were a number of reasons for this attitude: Gamma ray detectors had a resolution of about 5%. More precise measurements could be made only on low-energy gamma rays, using bent crystal spectrometers with a resolution of $\approx 10^{-4}$. More fundamental was the realization that gamma ray emission was accompanied by nuclear recoil so that the full energy of the nuclear transition was not available in the gamma ray, i.e. there was no possibility of resonant detection. Another limitation was recognized to arise in gaseous or liquid materials from the Doppler broadening due to the random thermal motion of the atoms. This leads to a width of $\approx 10^{-6}E$, where E is the energy of the gamma ray. The hope of resolving the hyperfine structure (hfs) splitting of nuclear levels, $\approx 10^{-10}E$, by measurements on nuclear gamma rays therefore seemed remote.

The picture changed drastically with the publication by Mössbauer (1958a,b, 1959) of his discovery of resonant scattering of nuclear gamma rays emitted and absorbed without recoil energy loss by the nuclei of atoms bound in solids. This work demonstrated that the theoretical limit of resolution, the natural linewidth determined by the decay time, could be directly achieved under certain circumstances. For isotopes which have since been used in Mössbauer effect (ME) experiments, this corresponds to a theoretical resolution of $10^{-10}E$ to $10^{-15}E$. This discovery opened up a new area of research, gamma ray hfs spectroscopy.

The concepts required to understand Mössbauer's discovery are basic

1

and simple *. It is only necessary to realize that a crystalline solid is a mechanical system whose vibrational properties must be described in quantum mechanical terms. The probability for a change in the quantum state of such a system by an impulse excitation is readily computed. In terms of the Debye model of lattice vibrations it turns out that the probability for excitation vanishes rapidly when the energy available for excitation becomes less than the Debye energy $k\theta_D$. The energy available to excite the crystal, in the case under consideration, is the nuclear recoil energy E_R due to gamma ray emission. Computed for a free atom of mass M, this energy is

$$E_R = \frac{E^2}{2Mc^2},\tag{1.1}$$

where c is the velocity of light. The probability for no change in the quantum state of the lattice, also called the recoil-free fraction, is

$$f = \exp\left[-\frac{E^2}{2Mc^2k\theta_D}\left(\frac{3}{2} + \frac{\pi^2T^2}{\theta_D^2}\right)\right], \quad T \ll \theta_D.\tag{1.2}$$

In the limit of low temperature, $T \ll \theta_D$, and small recoil energy, $E_R < k\theta_D$, the probability of lattice excitation is simply $3E_R/(2k\theta_D)$. For a number of isotopes and materials used, this factor is less than 0.1.

The existence of events in which the lattice is not excited, so-called zero-phonon events, forms the basis for the understanding of the Mössbauer effect. The resulting gamma rays have the full energy of the nuclear transition, and therefore the proper energy for resonant reabsorption. Their uncertainty in energy is determined by the width of the decaying nuclear level, i.e. its lifetime. The lifetimes of low-lying nuclear states are generally in the 10^{-6} to 10^{-9} s range, implying widths of 10^5 to 10^8 Hz, which are sufficiently small to make it possible to resolve nuclear hfs in favorable cases (fig. 1.1).

The widths of these gamma rays are so small that it is exceedingly unlikely that a gamma ray from a nuclear transition in one isotope can be resonantly absorbed by another. In practice this has never been observed. Mössbauer effect experiments are therefore invariably carried out with sources and absorbers utilizing the same nuclear transition. This is sometimes not apparent in the description of an experiment because the source may be described as containing the radioactive parent of the Mössbauer isotope. For example, experiments with the 14.4 keV transition between the ground and first excited state of ^{57}Fe utilize a source which contains ^{57}Co. The latter decays by

* More detailed treatments can be found in the items cited in the bibliography (p. 74).

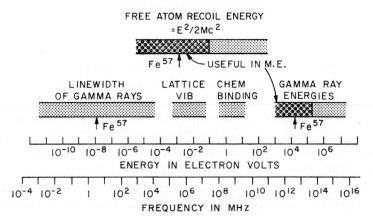

Fig. 1.1. Characteristic energies of solid state and nuclear properties pertinent to Möss-bauer effect. The horizontal bands denote characteristic ranges associated with the properties shown, but values outside the bands are not excluded. The 'free atom recoil energy' band corresponds to the range of the 'gamma ray energy' band and was computed for nuclei of atomic weight 100. The upper cut-off of the 'lattice vibration energy' band is based on Debye temperatures, the lower cut-off reflects the decreasing density of states of the Debye spectrum. Gamma ray linewidths are computed for characteristic lifetimes of nuclear excited states. The range of gamma ray energies in which the ME has been observed is shown by a distinctive cross-hatching. Values for the isotope ^{57}Fe are also indicated.

electron capture to an excited state of ^{57}Fe which then emits the Mössbauer gamma ray.

Experiments are usually carried out in transmission geometry similar to that originally used by Mössbauer. Scattering geometry has been found to be advantageous in only a few cases. In the earliest experiments (Mössbauer, 1958a,b), the existence of resonant absorption was shown by a substitution experiment utilizing a nonresonant absorber with equivalent electronic absorption properties. Subsequently Mössbauer showed that small Doppler velocities can be used to shift the energy of the gamma ray sufficiently to destroy the resonant absorption. In fact, carefully controlled Doppler velocities make it possible to map out the shape of the gamma ray line and to measure its hfs splitting. It is now common usage to talk about a 'Mössbauer effect spectrometer', in which the Doppler velocity performs a function related to that of the dispersive element in an optical absorption spectrometer. Note, however, that the two types of spectrometer operate on quite different principles. The Mössbauer spectrometer shifts the energy of an already narrow line, while the optical instrument disperses the radiation from a complex source.

For most experiments in which the subject of the investigation is the absorber, one chooses a source which emits a line having a width as close to the natural width as can be obtained. This generally calls for a cubic host environment to avoid broadening due to quadrupolar effects.

In the operation of an automatic ME spectrometer, the energy of the gamma ray of a radioactive source is Doppler modulated to sweep repeatedly across the region of resonant absorption of the absorber which contains nuclei of the same species. The gamma ray detector output is sorted into storage registers in a multichannel analyzer whose access is synchronized with the Doppler velocity of the source (fig. 1.2). The results are absorption

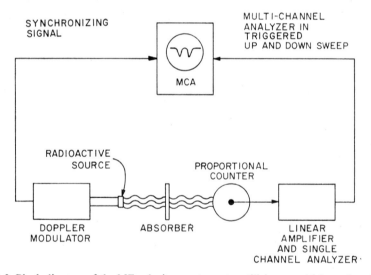

Fig. 1.2. Block diagram of the ME velocity spectrometer utilizing a multichannel analyzer (MCA) to accumulate an absorption spectrum. The energy of the gamma rays from the source is Doppler modulated in synchronism with the channel address of the MCA.

spectra like those shown in fig. 1.3, in which the energy scale is usually represented by the Doppler velocity. The relation of Doppler velocity v to energy shift δE is

$$\frac{\delta E}{E} = \frac{v}{c}. \tag{1.3}$$

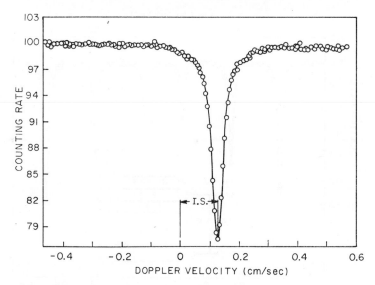

Fig. 1.3. Typical ME absorption spectra for ^{57}Fe, obtained with a source which emits an unsplit ^{57}Fe gamma ray line, i.e. ^{57}Co in a Pd metal host.
(a). Paramagnetic ion in a cubic site, RbFeF$_3$ at 127 K. E = 1.44 × 10^4 eV, I.S. = 6.14 × 10^{-8} eV.

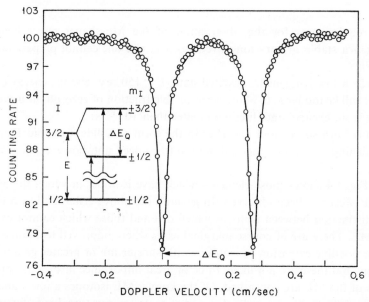

Fig. 1.3(b). Paramagnetic ion in a noncubic site, FeF$_2$ at 78.4 K. E = 1.44 × 10^4 eV, Δ E$_Q$ = 1.40 × 10^{-7} eV.

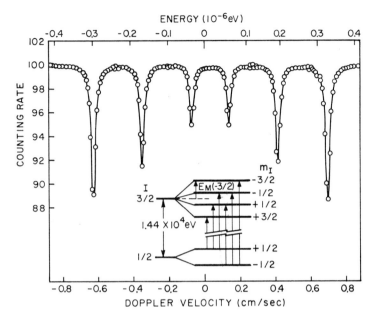

Fig. 1.3(c). Iron in a magnetically ordered compound, FeF$_3$, at 291 K.

The requirements for the observation of the Mössbauer effect include:

(1) a stable or very long lived isotope, else one cannot prepare an absorber;

(2) a low-lying nuclear excited state, $E < 150$ keV, else the energy of the recoil will be too large to allow a reasonable fraction of zero-phonon events;

(3) an excited state lifetime greater than 10^{-13} s;

(4) both source and absorber in the form of solids, to avoid Doppler broadening by random thermal motion (they need not, however, be crystalline).

Fig. 1.4 shows those elements which have isotopes in which the Mössbauer effect has been observed. In general, the above criteria allow a clearcut distinction between isotopes which can and those which cannot exhibit the ME. There are of course marginal cases where improved technique may make possible extension to isotopes with shorter life or greater gamma ray energy, but these are not likely to be of great importance insofar as applications of the ME are concerned. Of the known ME isotopes a few stand out because the effect is so easily observed at room temperature due to favorable recoil-free fractions and linewidths. These include ^{57}Fe, ^{119}Sn, ^{129}I, ^{161}Dy

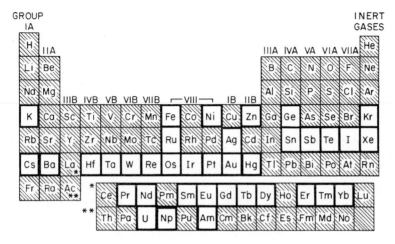

Fig. 1.4. Periodic table showing those elements in which the ME has been observed. (Adapted from 'Mössbauer Effect Data Index' by Muir et al., 1966.)

and ^{151}Eu. Many other isotopes require low temperature to obtain a good recoil-free fraction.

Mössbauer spectra contain information of interest to both nuclear and solid state physics in (a) the linewidth, (b) the strength of the absorption, (c) the shift of its centroid and (d) the hfs splitting.

(a). The minimum attainable linewidth is the natural width $\Gamma_n = \hbar/\tau$ determined by the decay time τ of the nuclear level. Experimentally obtained linewidths are usually somewhat larger because of finite absorption and geometric effects which can be accurately computed and are not of fundamental import. Broadening of Mössbauer lines in paramagnetic ions can also arise if the electron spin relaxation time is sufficiently long so that the hyperfine interaction is not averaged out in the Larmor period. In fact it has been possible to observe the whole range of relaxation time, from that corresponding to a narrow paramagnetic line to fully resolved hfs, by varying the temperature of a single compound (Wickman and Wagner, 1969). The linewidth may thus contain information on the nuclear decay time and on electron spin relaxation. In certain favorable cases it may also be affected by crystalline defects which manifest themselves through inhomogeneous quadrupolar or isomer shift effects.

(b). The strength of the absorption is measured by the cross section σ which is a product of the cross section for nuclear resonant aborption σ_0 and the recoil-free fraction f. The former is given by

$$\sigma_0 = \frac{\lambda^2}{2\pi} \frac{2I_e + 1}{2I_g + 1} \frac{1}{1 + \alpha},$$ (1.4)

where $\lambda = hc/E$ is the wavelength of the gamma radiation, I_e and I_g the nuclear spins of the excited and ground state, respectively, and α the internal conversion coefficient. (For ^{57}Fe, λ is 0.86 Å.) The recoil-free fraction is given most generally by

$$f = \exp(-4\pi^2 \langle x^2 \rangle / \lambda^2),$$ (1.5)

where $\langle x^2 \rangle$ is the mean square displacement of the decaying nucleus in the direction of gamma ray emission. This expression reduces to eq. (1.2) if $\langle x^2 \rangle$ is evaluated in the Debye approximation. The nuclear parameters which enter the net absorption cross section are usually well known, so that measurements of the temperature dependence of f are generally used to obtain information on the vibrational properties of the lattice or impurity atoms bound in solids.

(c). The precise value of the energy of a Mössbauer gamma ray transition depends both on temperature and on the chemical environment of the emitting or absorbing nucleus. This manifests itself as a shift of the centroid of the ME absorption from zero Doppler velocity when source and absorber are chemically distinct or are at different temperatures (fig. 1.3a). Two mechanisms are involved. The first is the relativistic second-order Doppler effect due to thermal motion *; the second is the electrostatic interaction of nuclear and electronic charge.

The second-order Doppler shift, also called thermal red shift, is given most generally by

$$\frac{\delta E}{E} = -\frac{\langle v^2 \rangle}{2c^2} = -\frac{U}{2c^2},$$ (1.6)

where $\langle v^2 \rangle$ is the mean square velocity of the emitter atom and U is the internal energy per unit mass. Note that the fractional shift is equal to the ratio of kinetic energy to relativistic rest energy, which is $\approx 10^{-12}$ at room temperature.

* The existence of a *second*-order kinematic shift raises the question of why there is no *first*-order Doppler effect broadening in Mössbauer spectra. The conventional explanation is based on the notion that the average velocity of the emitting nucleus during its lifetime vanishes because lattice vibration periods are very much smaller than the nuclear lifetime. This is indeed valid when one considers the recoil-free part alone, but if one considers the spectrum of all the emitted gamma rays, recoil-free plus phonon-accompanied, the first-order Doppler width is seen to exist.

In the Debye approximation the thermal shift δE_T is given by

$$\frac{\delta E_T}{E} = -\frac{3kT}{2Mc^2}\left[\frac{3}{8}\frac{\theta_D}{T} + 3\left(\frac{T}{\theta_D}\right)^3\int_0^{\theta_D/T}\frac{x^3 dx}{e^x - 1}\right]. \tag{1.7}$$

In the region where $T > \frac{1}{3}\theta$, it is well approximated by

$$\frac{\delta E_T}{E} = -\frac{3kT}{2Mc^2}\left(1 + \frac{\theta_D^2}{20T^2}\right). \tag{1.8}$$

At high temperature, the limiting value of the shift is

$$\frac{\delta E_T}{E} = -\frac{3kT}{2Mc^2}, \tag{1.9}$$

which is 7.25×10^{-5} cm·s^{-1}·K^{-1} for ^{57}Fe. This is sufficiently large to be readily observed in ME spectra.

The second contribution to the centroid shift arises from the electrostatic interaction of nuclear and electronic charges. In general, the effect is due to the s-electron charge density $|\psi(0)|^2$ at the nucleus and to the difference $\delta\langle r^2\rangle$ between the mean square nuclear charge radii in the isomeric excited and ground states. Because of this, it is generally called isomer shift or sometimes chemical isomer shift, here abbreviated IS. It is given by the equation

$$IS = \tfrac{2}{3}\pi Ze^2 \left(|\psi_a(0)|^2 - |\psi_s(0)|^2\right)\delta\langle r^2\rangle, \tag{1.10}$$

where subscripts a and s stand for absorber and source. In order to facilitate intercomparison, IS's should be expressed with respect to a common reference substance. In this volume we have occasionally converted ^{57}Fe measurements to a metallic iron reference in order to make results from various sources intercomparable. We will show the reference substance as a subscript, e.g. IS_{Fe}.

In reading the literature one finds considerable ambiguity regarding the sign of the shift. An (unfortunate) convention calls a shift positive whenever the Doppler motion is such that source and absorber approach each other. This leads to the artifact that the shifts of identical states in source and absorber have opposite signs. We shall define *positive* Doppler velocity as corresponding to an *increase* in energy.

Isomer shifts provide information particularly useful to chemists because they are sensitive to valence and covalency, which affect s-electron charge density. They can be used to identify the chemical state of isolated impurity atoms and submicroscopic precipitates.

(d). The ability to resolve the hfs splitting of nuclear levels is the most useful application of the ME. It competes here with conventional magnetic resonance but provides additional information, and can be used in certain cases where nmr and epr fail. One major difference is that one always obtains the hfs of both ground and excited nuclear states. This makes it possible to determine the nuclear moments of short lived excited states not accessible by nmr. It also means that measurements can be made in cases where the ground state nuclear spins are zero.

The hfs spectra studied by the ME show the effects of the electric qua-drupole (fig. 1.3b) and magnetic dipole (fig. 1.3c) interactions. When a magnetic field H acts on a nucleus with spin I, the $2I+1$-fold degeneracy is lifted according to

$$E = -\mu H m_I / I, \qquad\qquad -I < m_I < I, \qquad\qquad (1.11)$$

where μ is the nuclear magnetic moment. In magnetically ordered systems, the hfs splitting which arises largely from core and conduction electron polarization is often given as an effective magnetic field H_{eff}.

When a nucleus with $I > \frac{1}{2}$ is in an axially symmetric electric field gradient, the quadrupole interaction results in only a partial lifting of the degeneracy, according to

$$E_Q = \frac{e^2 q Q}{4I(2I-1)} [3m_I^2 - I(I+1)], \qquad\qquad (1.12)$$

where E_Q is the quadrupolar energy shift, eq is the major component of the electric field gradient (EFG) tensor, Q the nuclear electric quadrupole moment.

If both a magnetic field and an EFG act on a nucleus with $I > \frac{1}{2}$, the details of the splitting depend on the orientation of the magnetic field with respect to the principal axes of the EFG tensor. Simple solutions exist only for H along a principal axis; numerical values for other cases have been tabulated.

In the following sections, we shall examine how information concerning crystalline defects has been obtained from Mössbauer spectroscopy. This application of the ME was pointed out by Gonser and Wiedersich (1963) while the field was still in its early development. We will take a generous view of the topics which fall into this category and include radiation damage, vacancy association, order–disorder transformations, structural defects such as antiphase domain boundaries and precipitation of second phase, and chemical impurities both from the point of view of their effects on the host and the study of their own properties. We will also discuss briefly the detection of defects produced by the radioactive decay which precedes the emission of the ME gamma ray in most isotopes.

References

Gonser, U. and H. Wiedersich, 1963, in: Proc. Intern. Conf. on Crystal lattice defects, 1962, Conf. J. Phys. Soc. Japan 18, Suppl. II, 47.

Mössbauer, R. L., 1958a, Z. Physik **151**, 124.

Mössbauer, R. L., 1958b, Naturwissenschaften **45**, 438.

Mössbauer, R. L., 1959, Z. Naturforsch., **14a**, 211.

Wickman, H. H. and C. F. Wagner, 1969, J. Chem. Phys. **51**, 435.

2 | LATTICE DEFECTS

2.1. Radiation damage

Radiation effects in solids arise from ionization and displacement of atoms. In monatomic solids, permanent damage arises only from atomic displacement. Near the threshold energy for displacement a vacancy-interstitial pair is produced. The nature of the damage found some finite time after the displacement process depends on the mobility of these two elementary defects and their tendency to annihilate, cluster or associate with chemical impurities. At higher energy, the primary knock-on initiates a displacement cascade whose final stage has sometimes been thought of as a thermal spike. The net result of such a cascade is a group of spatially correlated vacancies and interstitials. An important feature of such displacements is the replacement collision in which an energetic atom ejects a second atom from a lattice site and itself remains in it. In more complex solids, e.g. intermetallic compounds, such an event disorders the lattice.

In principle, the Mössbauer effect provides a technique which is uniquely adapted to the study of these defects. The radioactive parent of the Mössbauer isotope is normally produced by a nuclear reaction, neutron capture or Coulomb excitation. In each of the processes, sufficient kinetic energy is imparted to the atom to cause it to be displaced from its lattice site. The resulting displacement cascade takes place in a time short compared to the lifetime of levels useful in ME studies, and even thermal after-effects are largely if not entirely dissipated before the gamma ray is emitted. Since the gamma ray is emitted by the atom which initiated the damage process, one might expect that the ME would be much more sensitive to radiation damage

12

in experiments in which the *source* is prepared by nuclear reaction or ion implantation than in experiments using irradiated *absorbers*.

In most of the experiments done in the early years of ME research, steps were taken to avoid radiation damage, e.g. sources were annealed or prepared by diffusion of radioactive parents into undamaged host crystals. The first indications of radiation damage effects were reported in ME studies of rare earth isotopes, in which it was found that the linewidth of sources prepared by irradiation of the oxide could be improved by annealing (Wiedemann et al., 1963). It is surprising then, to find that radiation damage effects have turned out to be quite elusive in cases where they were fully expected, for example in ion implantation or Coulomb excitation ME work.

One of the earliest definitive ME radiation damage experiments was carried out using thermal neutron capture in* $Mg_2{}^{118}SnO_4$, an inverse spinel in which tin is normally present as Sn^{4+} (Hannaford et al., 1965; Hannaford, 1968; Hannaford and Wignall, 1969). A significant fraction of the ^{119m}Sn produced in these spinel *sources* was present as Sn^{2+} in a noncubic lattice site with a small, anisotropic recoil-free fraction, as can be seen in fig. 2.1. These facts, as well as the additional finding that the fraction of Sn^{2+} is decreased when reground and refired material is used, suggest that the divalent tin is stabilized by lattice defects. It was shown that the observed effect is not due to bulk damage by demonstrating that a $Mg_2{}^{119}SnO_4$ *absorber* remained unchanged by reactor irradiation.

The appearance of the resolved subsidiary line was attributed to a defect configuration consisting of a divalent ^{119m}Sn ion in an octahedral site associated with a charge-compensating oxygen vacancy (fig. 2.2). The recovery of these irradiation effects was found to take place in a single annealing stage between about 600 and 900 °C. The annealing data for this stage indicate that the defects are distributed over a narrow band of activation energies in the range 3–4 eV which may correspond to the energy required to dissociate the Sn^{2+}–oxygen vacancy defect complex.

The production of this defect seems to be confined to Mg_2SnO_4 and (to a lesser extent) $ZnMgSnO_4$. There was no evidence of similar irradiation effects in other spinels, i.e., Zn_2SnO_4, $MgNiSnO_4$, $Mg_{1.5}FeSn_{0.5}O_4$, $MgAl_2(4\% \ Sn)O_4$, $Mg_2Ti(10\% \ Sn)O_4$; in perovskites, e.g., $BaSnO_3$, $SrSnO_3$, $CaSnO_3$ or Li_2SnO_3; or in SnO_2. Moreover, the presence of only fractional amounts of cation impurities (e.g., 0.25 Ni^{2+}, Ti^{4+} or Fe^{3+}) in the Mg_2SnO_4 lattice was sufficient to reduce the observed defect concentration very markedly. This suppression was tentatively attributed to a

* Decay schemes for some common Mössbauer isotopes are given in appendix 2.

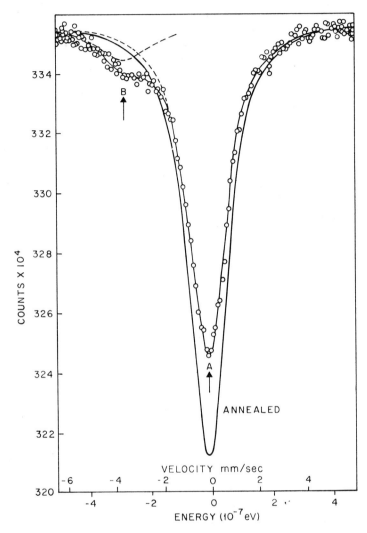

Fig. 2.1. Neutron irradiation damage in the ^{119}Sn Mössbauer spectrum of a $Mg_2{}^{119m}SnO_4$ source. The absorber is SnO_2. Line A corresponds to Sn^{4+}, as in annealed $MgSnO_4$, line B to Sn^{2+}. (From Hannaford et al., 1965.)

reduction in the stability of the defects or in the population of the Sn^{2+} level. The former may arise from an increase in the fraction of admixed 5p-wavefunction on the Sn^{2+}–oxygen vacancy complex. The latter may arise primarily from a decrease in the oxygen u-parameter which will raise the

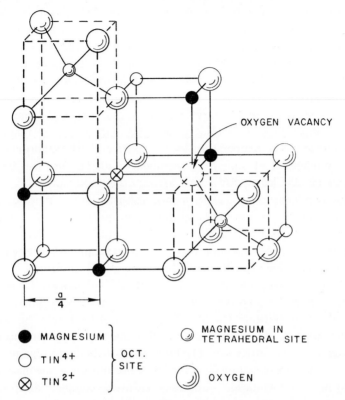

OXYGEN VACANCY

$\frac{a}{4}$

● MAGNESIUM ⎫
○ TIN⁴⁺   ⎬ OCT. SITE
⊗ TIN²⁺   ⎭

MAGNESIUM IN TETRAHEDRAL SITE

OXYGEN

Fig. 2.2. Part of the unit cell of the spinel structure showing the defect proposed for Mg_2SnO_4.

energy of the Sn^{2+} level relative to the oxygen 2p-band at the octahedral sites.

Radiation effects have also been found in SnO by Bondarevskiy and Seregin (1968). They show that reactor irradiation converts an appreciable fraction of the tin to Sn^{4+}. They propose that this is due to the displacement of tin atoms from normal lattice sites with the formation of F centers. The fast neutron flux which is responsible for this effect was unfortunately not given in the paper. They also show that ^{119m}Sn atoms produced by thermal neutron capture in ^{118}SnO are dominantly in normal SnO lattice sites, although the sample was exposed to an even greater fast neutron flux. It is not clear whether the absence of Sn^{4+} is due to differences in the chemical state of the two SnO samples or to radiation annealing.

In a similar experiment, Stepanov and Aleksandrov (1967) reported that the ^{125m}Te emission line from neutron irradiated PbTe has an isomer shift relative to an unirradiated PbTe absorber. This shift anneals at room temperature with a lifetime of ≈ 10 d. The nature of the damage was not defined. More recently, Ullrich and Vincent (1969) in experiments on PbTe, Te and TeO$_2$ report no measurable changes in line width or position after thermal neutron capture at 45 °C. They suggest that the results of Stepanov and Aleksandrov are due to fast neutron damage which was minimized in their own experiment. It is also conceivable that the pile irradiation was carried out at lower temperature by Stepanov and Aleksandrov resulting in greater retention of damage. Experiments with fast neutron irradiated absorbers should clearly distinguish between these alternatives.

Closely related to these experiments, at least from the point of view of radiation damage, are ion implantation experiments in which a long-lived ME parent isotope is injected into a host crystal using a mass analyzer. The damage due to the implantation process is in the immediate environment of the implanted ion. De Waard and Drentje (1966) implanted ^{129}Te into metallic iron to study the hfs splitting of the ^{129}I daughter. Their spectra show the presence of two types of iodine, one with large magnetic hfs, the other with small splitting. The hfs effective field of the former fits well into the systematics of H_{eff} of neighboring elements in dilute solutions in iron indicating a normal lattice site. The latter was attributed to regions of high ^{128}Te concentration resulting from the limited resolution of the mass separator used in the implantation. Annealing led to the complete disappearance of the resolved hfs spectrum, presumably due to the reaction of Te with Fe to form iron telluride. This finding precludes more detailed study of possible implantation damage effects in this case. In a subsequent experiment, de Waard et al. (1968) prepared sources which exhibited only the spectrum with large hfs and concluded that the implanted ions were in substitutional sites.

Similar implantation experiments have also been carried out with ^{129}Xe (Nilsen et al., 1967). The data obtained in these cases are not sufficiently detailed to allow any conclusions regarding the defect environment of the implanted ion.

The time scale between defect production and emission of the ME gamma ray in the experiments just discussed is of the order of hours or days. During that time significant annealing may have taken place. This is certainly true in metals in which interstitials or vacancies are mobile at room temperature. The time scale can be greatly reduced by observing the ME of gamma rays following alpha decay or during in-beam excitation.

The Mössbauer effect after alpha decay was first observed in ^{237}Np following the decay of ^{241}Am in NpO_2 by Stone and Pillinger (1964). The excited state has a half-life of 6.3×10^{-8} s. Comparison with data taken with a beta decay ^{237}U source and the same absorber show a significantly reduced effective recoil-free fraction in the case of alpha decay. The reduced f could be attributed to displacement to a variety of sites with distinct isomer shifts and hfs structures or to local heating (Mullen, 1965).

Subsequently, experiments with ^{241}Am alpha decay sources were carried out by Dunlap et al. (1968) who found that metallic sources, especially in thorium and americium hosts, give a narrow linewidth and a good recoil-free fraction. (The best linewidth obtained is, however, still 15 times the natural width.) No indications of displaced atoms were found in metallic sources, but multiple emission lines were reported for AmO_2 sources. The broadened line and reduced f reported by Stone and Pillinger (1964) should then probably be attributed to the existence of these multiple lines in their oxide sources rather than to thermal after-effects.

Additional radiation damage results became available with the advent of in-beam Mössbauer studies utilizing such techniques as recoil ion implantation, nuclear reaction, Coulomb excitation and neutron capture. The first of these experiments were done with ^{40}K, an isotope for which a long-lived parent is not available. Ruby and Holland (1965) obtained a weak ME using a potassium target excited by a (d,p) reaction and a KCl absorber. Hafemeister and Shera (1965), in more detailed experiments using neutron capture, found that radiation damage did not substantially diminish the recoil-free fraction in KCl, KF or K metal. The maximum recoil energy imparted to the K atom by the neutron capture gamma rays (800 eV) is quite sufficient to displace the atom from its original lattice site. The results suggest that the atom becomes thermalized in a normal site before the emission of the ME gamma ray. The half-life of the 29.4 keV first excited state of ^{40}K is 3.9×10^{-9} s.

Coulomb excitation experiments with ^{61}Ni (Seyboth et al., 1965), show no significant change in f but some line broadening compared to experiments with radioactive sources. Similar experiments with ^{73}Ge (Czjzek et al. 1966, 1967; Zimmerman et al., 1968) have been carried out by implanting the recoil into metallic chromium as well as into germanium itself. In germanium, the effective recoil-free fraction was quite small; X-ray crystallographic examination showed that the target had become amorphous during the bombardment.

A large number of rare earth isotopes not accessible by radioactive decay have also been studied by Coulomb excitation, e.g., ^{172}Yb, ^{174}Yb and ^{176}Yb (Eck et al., 1967a,b), ^{160}Gd, ^{164}Dy and ^{168}Er (Stevens et al., 1967). Rare earth sesquioxide or metallic targets were used. No radiation damage effects were observed. Fink and Kienle (1965), using neutron capture to study ^{156}Gd and ^{158}Gd, found a slightly broader line with a Gd_2O_3 target than with a metallic target. They concluded that recoil effects produce only small differences in the hfs between a metallic and an oxide target. On the other hand, Jacobs et al. (1969) found definite evidence for radiation damage in HFC and HFN associated with the recoil of excited nuclei following Coulomb excitation.

A number of sensitive experiments have been done with ^{57}Fe using various forms of excitation. The ME of ^{57}Fe in metallic iron following Coulomb excitation by 3 MeV alpha particles was first demonstrated by Lee et al. (1965). In a more detailed report, Ritter et al. (1967) show that there is no difference in recoil-free fraction, isomer shift or hyperfine effective field compared to those of a metallic iron absorber, indicating that the excited atoms return to normal lattice sites. In α-Fe_2O_3 they report a reduction in the effective recoil-free fraction. This is probably due to displaced atoms which do not contribute to the normal hfs lines, rather than to an actual change in f for atoms on normal lattice sites.

Similar experiments using the (d,p) reaction on ^{56}Fe to populate the ^{57}Fe excited state were reported by Goldberg et al. (1965) in metallic iron, and by Christiansen et al. (1966) in stainless steel. In both cases the normal ME was found. In a more detailed account, Christiansen et al. (1967) show that there is no indication of any lattice perturbation in the vicinity of the excited recoil atom, i.e. there is no indication of atoms in interstitial sites or atoms with nearby defects.

Experiments using neutron capture in ^{56}Fe were reported by Berger et al. (1967) and Berger (1969). The spectrum of metallic iron was again identical to that of an iron absorber with regard to isomer shift, quadrupole splitting and hyperfine field. In Fe_2O_3, the hfs effective field was 4% smaller than that in an absorber, but other aspects were unchanged. In $FeSO_4 \cdot 7H_2O$ about 40% of the iron was found to be trivalent. This interesting result is discussed in section 2.3.

From the point of view of radiation damage, the in-beam experiments may be criticized on the ground that the target is being progressively damaged by the beam as the experiment proceeds. This objection is overcome by recoil implantation in which the excited atom is ejected from the target and implanted into a catcher foil which is not subject to irradiation. In such experiments, Sprouse et al. (1967) found that Coulomb-excited ^{57}Fe recoil implanted in copper, aluminum, gold and iron were indistinguishable from ^{57}Fe produced by the decay of ^{57}Co. Implantation into $(Fe,Mg)_2SiO_4$ (olivine) yielded a greatly reduced effect.

The common finding of all these experiments is that in a simple metal the primary displaced or implanted atom returns to a normal lattice site. This result is in line with theoretical calculations which showed the importance of replacement collisions for primaries with energies up to 100 eV (Gibson et al., 1960; Vineyard, 1961, 1963; Erginsoy et al., 1964, 1965; see also Goldberg et al., 1966; Dedericks et al., 1965).

Experiments in oxides and salts have not been done in sufficient detail to allow an unambiguous interpretation of the data. Thus, it is generally not known whether the reduced ME which is often found is due to a small recoil-free fraction or to a sizable fraction of atoms in other than normal lattice sites.

It is apparent from this survey that whereas radiation effects have been occasionally observed, there is as yet little detailed interpretation in the published ME literature. The reason may well be that in most cases where such effects could have been studied, steps were taken to avoid them because they would have interfered with the primary objectives of the experiment. Nevertheless, there is sufficient indication that such effects are detectable and could be studied if there were sufficient motivation. The recognition that replacement collisions are responsible for the absence of damage effects in simple metals is one of the important findings for those experiments where damage effects are to be minimized.

Replacement collisions in intermetallic compounds will produce disorder which can be studied by the ME. Such effects have been demonstrated by Berger (1969) in FeAl and Fe_3Al sources using gamma recoil after thermal neutron capture. They could also be studied in absorbers subjected to fast neutron damage.

2.2. Plastic deformation

A careful comparison of the ^{57}Fe hfs of cold rolled and annealed iron foil absorbers has been made by Cohen (1970). He found no measurable difference in the room temperature absorption spectra indicating that the dislocations affect at most a very small fraction of the iron atoms. Such effects have, however, been seen in the absorption spectrum of ^{181}Ta in plastically deformed Ta foils (Sauer, 1969). The natural linewidth of the ^{181}Ta 6.2 keV gamma ray is about 70 times smaller than that of ^{57}Fe. The

ME of [181]Ta is therefore very sensitive to external perturbation, so much so that it has proved very difficult to observe. Sauer has achieved a linewidth about 9 times the natural width using carefully prepared [181]W in W metal sources and Ta metal foil absorbers.

Upon plastic deformation, an increase in linewidth by a factor of three was recorded. This change was found to saturate after a sample elongation of about 5%. Plastic deformation did not change the IS indicating that the line broadening must be of quadrupolar origin. The broadening could be entirely removed by subsequent annealing.

Line broadening and an increase in the IS were also observed upon the introduction of oxygen or nitrogen into the lattice. Both were found to correlate linearly with the change in lattice constant produced by the impurity.

Other experiments dealing with plastic deformation are discussed in chapter 4. These are concerned not with lattice defects but with stress-induced crystallographic transformation and with slip-induced disorder in intermetallic compounds.

2.3. Ionization effects

The ionizing effects of nuclear radiation do not produce permanent changes in most solids. The chief exceptions are the generation of color centers in ionic crystals and the radiolysis of molecular crystals. In metals and semiconductors, ionization does not produce structural changes. Varley (1954a,b) has suggested a mechanism by which ionization could produce atomic displacement in ionic salts, but defect production by this mechanism has not been demonstrated experimentally.

In ME experiments using a radioactive parent isotope in a nonmetallic host, highly ionizing radiation emitted just prior to the emission of the Mössbauer gamma ray may produce various types of damage. This includes not only alpha and beta decay parents but also electron capture and internal conversion isotopes which emit Auger electrons. These produce radiolytic effects in the immediate vicinity of the Mössbauer atom.

There are a number of other effects which can arise from the loss of electrons by the Mössbauer atom itself during the Auger cascade. Consider the consequences of a nuclear decay process which removes an inner shell electron, e.g. electron capture or internal conversion. It is well established that the filling of the resulting hole is accomplished largely by an Auger cascade in which electrons with X-ray energies are emitted by the atom, leaving it in a highly charged state (Siegbahn, 1965). This process requires no more than $\approx 10^{-14}$ s. In a solid, it is followed by what has been called charge relaxation, i.e. the recapture of electrons from the filled bands or surrounding ligands to return the atom to a 'normal' valence state. The effects of these violent events on the chemical bonding of the atom and its neighbors are not readily predicted. Among the conceivable consequences are the creation of lattice defects and Coulombic disruption of the original molecule (Carlson and White, 1965). Charge relaxation may then proceed to valence states differing from the initial one, or even to atomic excited states. It is clear that the nature of the host crystal will have an important effect on the type of transformation which takes place.

We next consider the experimental evidence which bears on this subject. Radiolysis by an external source of ionizing radiation has been studied in a number of ME experiments. Saito et al. (1965) demonstrated the radiolysis of ferric oxalate ($Fe_2(C_2O_4)_3 \cdot xH_2O$) by ^{60}Co gamma rays. About one half of their material was converted to the ferrous state ($FeC_2O_4 \cdot xH_2O$) by a dose of 200 Mrad. The radiolytic decomposition and thermal annealing of $K_3Fe(C_2O_4)_3 \cdot 3H_2O$ as well as of the anhydrous compound has been studied in detail by Dharmawardena and Bancroft (1968) and by Temperley and Pumplin (1969). The former show that both compounds decompose to an octahedral Fe(II) compound via a thermally unstable Fe(II) intermediate which is four-coordinated in the hydrated but six-coordinated in the anhydrous form. During thermal annealing, some of the unstable Fe(II) intermediate reverts to the initial Fe(III) compound while the remainder is converted to the octahedral Fe(II) compound. The latter identify the products of radiolysis as *cis*- and *trans*-isomers of potassium bisoxalato bisaquo ferrate(II). Evidence for three additional compounds, including potassium bisoxalato bisaquo ferrate(III) was also found in irradiated potassium trisoxalato ferrate(III) or in synthesized potassium–iron–oxalate complexes. The reduction of the Fe(III) is due to the radiolysis of the oxalate ion with the liberation of CO_2. The analogous radiolytic reduction of ferric citrate has also been observed by the ME (Buchanan, 1970).

In a related experiment, it was found by Sano and Hashimoto (1965) that the decay of ^{57}Co in cobalt(II) oxalate produces only Fe(II) oxalate. This result is entirely in line with the radiolysis experiments above since the Auger electron ionization has no tendency to oxidize ferrous oxalate. In a similar experiment, Sano and Kanno (1969) report the reduction of Sn(IV) to Sn(II) in $K_6Sn_2(C_2O_4)_7 \cdot 4H_2O$ sources containing the ^{119}Sn isomer. In this case the Auger electrons responsible for the radiolysis are the result of the internal conversion decay of the ^{119m}Sn.

Results indicative of chemical changes in the environment following electron capture decay had previously been obtained in a number of molecular crystals. In cobalt(III) acetylacetonate (Co(acac)₃), Wertheim et al. (1962) reported both Fe(II) and Fe(III) after the decay of $^{57}Co(III)$. The compound was synthesized from cobalt chloride containing the radioisotope and was repeatedly recrystallized. The trivalent iron had a spectrum like that of Fe(III) acetylacetonate, the Fe(II) was identified by its IS. The mechanism proposed for the production of Fe(II) was based on the modification of the ligand structure by the Auger electron ionization and subsequent charge relaxation by the iron ion to a state appropriate to the new ligand structure.

More recently, Hazony and Herber (1969) have suggested that the production of Fe(II) is due to an 'internal pressure' on the Fe(III) ion. This is based on the finding that in solids at high pressure Fe(III) reduces reversibly to Fe(II) (Drickamer et al., 1969). The 'internal pressure' is postulated on the basis of the density difference between Co(acac)₃ and Fe(acac)₃ even though the ionic radii of Co(III) and Fe(III) differ only slightly. The similarity of the spectra of Fe(acac)₃ under pressure as an absorber and $^{57}Co(acac)_3$ as a source support this idea.

Effects similar to those in Co(acac)₃ were also noted by Wertheim and Herber (1963) in cobalticinium tetraphenyl borate. Recently these experiments have been extended to other cobalt chelates. Nath et al. (1968a, 1970) report reduction to Fe^{2+} in one-fourth of the events following the decay of ^{57}Co in tris-dipyridyl cobalt(III) perchlorate. In bis-indenyl cobalt(III) perchlorate the emission spectrum bears no resemblance to that of the corresponding Fe(II) or Fe(III) complexes, leading to the suggestion that the molecule has been fragmented by the electrostatic repulsion resulting from the withdrawal of electrons from the ligands by the charge relaxation following the Auger cascade. On the other hand, in complex molecules with conjugated ring systems, such as vitamin B_{12}, its analogs and cobalt phthalocyanine, the same authors report that Mössbauer spectra characteristic of the entire molecule have been obtained following the electron-capture decay of ^{57}Co. This suggests that ^{57}Co emission spectroscopy may be useful for molecules of biological import. Evidence for molecular fragmentation has also been obtained in tris-phenanthroline cobalt(III) perchlorate by Jagannathan and Mathur (1969). They further suggest that the data on tris-dipyridyl cobalt perchlorate mentioned above can also be interpreted as showing evidence of such fragmentation, but occurring in a smaller fraction of events.

It is apparent that the study of these chemical transformations should be a fruitful area for further research.

In hydrated salts new valence states have been quite generally reported following electron capture decay. There is good agreement of the production of Fe^{3+} in ferrous ammonium sulphate hexahydrate by the decay of Co^{2+} (see for example Wertheim and Guggenheim, 1965; Triftshäuser and Craig, 1967) and in various cobalt compounds (Ingalls and de Pasquali, 1965; Ingalls et al., 1966; Friedt and Adloff, 1967, 1969). The latter have also examined the role of the water of hydration and find an increasing tendency toward the stabilization of defect charge states with increasing degree of hydration in $CoSO_4$. They agree with Mullen and Ok (1966) on the absence of Fe^{3+} in hydrated $CoCl_2$. Ingalls and de Pasquali (1965), on the other hand, show that Fe^{3+} is resolved at 143 K, while their room temperature data agree with those of the other two groups. The results suggest that the stabilization of Fe^{3+} in hydrated salts depends on temperature, or that the Fe^{3+} has a much lower effective Debye temperature than Fe^{2+}. The notion that the Fe^{3+} has a temperature dependent decay time has been ruled out by delayed coincidence ME experiments in which the spectrum was examined as a function of time during the decay of the excited state (Triftshäuser and Craig, 1966, 1967). These experiments gave no indication of any charge relaxation during the accessible time, i.e. from 0.2τ to 1.7τ, in compounds containing ^{57}Co, including $Fe(SO_4)_2(NH_4)_2 \cdot 6H_2O$, which show more than one valence state. Sample preparation was sufficiently reliable so that there can be no question but that the ^{57}Co was all in a single valence state prior to decay. The most likely conclusion is that charge relaxation goes to completion in a time short compared to the nuclear lifetime, but that there are a number of different final states corresponding to changes in the environment of the ion. Less likely is the possibility that charge relaxation is slow compared to the nuclear lifetime and that the final state is not reached in the time accessible in this experiment.

Friedt and Adloff (1967, 1969) have suggested that the environmental change responsible for the stabilization of higher charge states in hydrated compounds is due to the radiolysis of water. This has recently been followed up by Gütlich et al. (1968) and Wertheim and Buchanan (1969). The latter show that ^{60}Co gamma irradiation or high energy bombardment of ferrous ammonium sulphate hexahydrate produces a material whose Mössbauer absorption spectrum is just like the emission spectrum of the ^{57}Co doped compound (fig. 2.3). It was also shown that the Auger electron ionization in the molecule containing the ^{57}Co atom was sufficient to radiolyze a

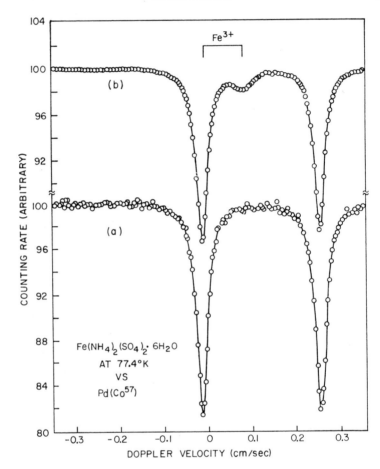

Fig. 2.3. Mössbauer absorption spectra of $Fe(NH_4)_2(SO_4)_2 \cdot 6H_2O$ (a) before and (b) after irradiation with 430 Mrad of 1 MeV electrons from a Van de Graaff accelerator. The Fe^{3+} contribution in the line near zero velocity in (b) was determined by using the known intensity and position of the Fe^{2+} line. (From Wertheim and Buchanan, 1969.)

water molecule in about $\frac{1}{3}$ of the decay events, a result in good agreement with the experimental yield of Fe^{3+}. The oxidation of Fe^{2+} is thought to be due to the OH radical produced by radiolysis. Electron bombardment of anhydrous salts of Fe^{2+} did not produce any valence change. This mechanism also explains the Fe^{3+} reported by Berger et al. (1967) following neutron capture in $FeSO_4 \cdot 7H_2O$.

In anhydrous salts the evidence is at first sight contradictory. Valence states differing from that of the present isotope have been obtained in some cases. We shall here consider only those cases in which the radioactive parent atom has the same valence as the host lattice. Other cases, complicated by vacancy and defect association required for charge compensation, are discussed in chapter 3. Examples of complete charge relaxation include ^{57}Co in $CoCl_2$ (Mullen, 1965; Cavanagh, 1969), ZnF_2 (Wertheim and Guggenheim, 1965), and CoF_3 (Friedt and Adloff, 1969). Multiple charge states have been reported in $CoSO_4$ (Ingalls et al., 1966), CoF_2 (Friedt and Adloff, 1967, 1969; Cavanagh, 1969; Wertheim et al., 1969), and in $KCoF_3$ (Jagannathan et al., 1969).

The compounds ZnF_2 and CoF_2 in this brief list belong to a particularly interesting group of materials, the rutile structure fluorides, whose magnetic and optical properties have been widely studied. Another compound in this series, MnF_2, was recently used as a ^{57}Co doped source in a ME critical point experiment (Wertheim et al., 1968). It was mentioned there that only Fe^{2+} was detected following the decay of Co^{2+}. There is not necessarily any conflict between the results on ZnF_2 and MnF_2 on one hand and those on CoF_2 on the other, since there is a systematic variation in unit cell volume (table 2.1). The possible significance of this parameter was pointed out some years ago by Pollak (1962) who based his arguments mainly on relative ionic sizes. The results of a systematic investigation of the stabilization of Fe^{3+} in rutile structure fluorides by Wertheim et al. (1969) are also shown in table 2.1. There is a strong indication that the smaller Fe^{3+} ion is preferentially stabilized in a lattice site smaller than that of Fe^{2+} in FeF_2. This conclusion is not in contradiction to that of Hazony and Herber (1969), since the iron here is not in a ligand environment appropriate to Fe^{3+}.

TABLE 2.1

Fractions of Fe^{3+} produced at room temperature by the electron capture decay of $^{57}Co^{2+}$ in various rutile structure fluorides. The recoil-free fraction is assumed to be the same for Fe^{2+} and Fe^{3+}

Compound	Unit cell volume (Å^3)	Fraction of Fe^{3+} (%)
MnF_2	78.61	0
FeF_2	72.99	0
CoF_2	70.09	32
ZnF_2	69.32	1
NiF_2	66.69	59
MgF_2	65.13	82

These results do not imply that differences in ionic size alone are sufficient to account for the stabilization of a defect charge state. Wertheim et al.
(1968) assume that a charge-compensating defect, perhaps a trapped electron,
is also required. Cavanagh (1969) attributes the effect to a local distortion
of the crystal structure produced by Coulomb forces associated with the
short-lived high charge states. The nature of this distortion was not further
elucidated. Friedt and Adloff (1969a) asume that charge compensation is
provided by ionic vacancies produced by K-capture after-effects. In particular they propose a Co^{2+} and a F^- vacancy to maintain charge neutrality
near an Fe^{3+} ion. This suggestion has the advantage of accounting for the
decreasing fraction of Fe^{3+} at high temperature in terms of the thermal
annealing (or outward diffusion) of these vacancies. It might be noted that
a decrease in the fraction of the higher charge state with increasing temperature is a feature common to experiments in anhydrous salts, but not
to those in oxides mentioned below.

A great deal of experimental work has also been done in another series
of isostructural compounds, the NaCl structure oxides. The transition metal
members of this series present a more complicated picture, however, since
they are well known defect solids, see for example Kröger and Vink (1956)
or Verwey (1951). At high temperature, the composition is a function of both
temperature and oxygen partial pressure. From a microscopic point of view,
it is believed that this composition is realized in CoO through the introduction of cation vacancies (Fisher and Tannhauser, 1966). The highest
defect concentration which was obtained is 1.2% at 1350 °C in 1 atm of
oxygen. Concentrations as low as 0.01% are obtained at 1000 °C in an
oxygen partial pressure of 10^{-4} torr.

Multiple charge states of Fe have been clearly identified following the
decay of ^{57}Co in MgO by Chappert et al. (1967, 1969). They attribute the
presence of Fe^+ to charge compensation by substitional OH^-, and the
presence of Fe^{3+} to previous oxidation of the ^{57}Co. This point of view is
supported by the finding that the fractions of these valences are sensitive
to sample preparation. Triftshäuser and Schroeer (1969) examined the time
dependence of the charge states of Fe produced by the decay of ^{57}Co in
three oxides, including MgO. In no case were time-dependent intensities
actually found. However, an anomalous line-narrowing indicative of a
relaxation process was always present in the data with longer delay. The
narrowing clearly shows that the relatively broad lines usually found in these
spectra are not due to inhomogeneous broadening. It is not clear at present
whether this unusual effect is related to the stabilization of multiple charge

states or whether they are completely determined by pre-existing charge-compensating defects.

It has been established that both NiO and CoO can be prepared in which only Fe^{2+} is produced by the decay of $^{57}Co^{2+}$ (Bearden et al., 1964; Siegwarth, 1967; Ok and Mullen, 1968a). In MnO, Siegwarth did not succeed in preparing a pure Fe^{2+} source. The materials which exhibit only Fe^{2+} presumably correspond to oxides closest to stoichiometry.

Of more interest is the behavior of nonstoichiometric material which exhibits both di- and trivalent iron. In all three oxides the fraction of trivalent iron increases with temperature in a reversible way and greatly exceeds the fraction of excess oxygen. In CoO, this is clear indication that the trivalent iron is not due to the selective oxidation of the ^{57}Co. In NiO, all the iron is trivalent above 475 K (Bhide and Shenoy, 1966a; Ando et al., 1967). In CoO, the Fe^{2+} was hardly detectable above 800 K (Bhide and Shenoy, 1966b). This was confirmed by Triftshäuser and Craig (1967). In MnO, Fe^{3+} was detected at room temperature but not below the Néel temperature of 117 K (Siegwarth, 1967). The change in the Fe^{3+} fraction at low temperature in MnO and NiO, and the complete reversibility with temperature indicate that it is not due to oxidation of the parent cobalt, but represents the stabilization of a defect charge state by a mechanism not yet clearly established.

The mechanism for the temperature dependence in the terminal valence of iron produced by the electron capture decay of $^{57}Co^{2+}$ has not been elucidated. It has been suggested that charge relaxation may stop at Fe^{3+} in the vicinity of a cation vacancy. The temperature dependence may then be related to the enhanced diffusive motion of the vacancies at high temperature which favors association with the newly created chemical defects. Localized lattice heating by the preceding decay could also contribute to this effect. In a critical experiment, Trousdale and Craig (1967) have shown, however, that the change in trivalent fraction follows a change in temperature without a measurable time delay and therefore does not require diffusive atomic motion.

An alternate explanation by Ok and Mullen (1968a) postulates two types of CoO in which the decay-product iron is respectively entirely divalent and entirely trivalent. The first type corresponds to well known stoichiometric CoO, the second to a hypothetical phase containing 25% vacancies. They ascribe the presence of both valences to mixtures of these two phases. This proposal fails to explain the reversible, rapid change in Fe^{3+} fraction with temperature, as well as the fact that the Néel temperatures deduced from the Fe^{2+} and Fe^{3+} spectra agree closely. It has been shown by Schroeer and Triftshäuser (1968) that the properties of the second phase correspond to those of fine particles. These have been examined in NiO by Kündig et al. (1967). The interested reader should not fail to read the original articles, including the rebuttal by Ok and Mullen (1968b). In spite of the large effort which has gone into the study of CoO it is surprising to find no experiments on material prepared to have a definite composition according to the phase diagram of Fisher and Tannhauser (1966).

A number of other oxide systems have also been examined, e.g., Cu_2O and Al_2O_3 by Triftshäuser and Schroeer (1969), and $CoCr_2O_4$ (and $CoCr_2S_4$) by Jagannathan and Mathur (1969). Multiple valences were found in all cases. Work with cobalt compounds has the advantage that the valence of the ^{57}Co prior to decay can be safely assumed to be the same as that of the normal cobalt. In the case of the spinel $CoCr_2O_4$, the cobalt is divalent and occupies the cubic A-sites exclusively. The finding of only 25–28 % Fe^{2+} after the decay of Co^{2+} was attributed to the stabilization of higher charge states by cation vacancies. Unfortunately, no comparisons were made with ^{57}Fe doped $CoCr_2O_4$ absorbers. These might shed some light on the line broadening which was attributed to the cation vacancies.

Experiments of this general type have also been carried out with other isotopes. Klöckner et al. (1967) and Rother et al. (1969) report the detection of Ir^{2+}, Ir^{3+} and Ir^{4+} following the beta decay of ^{193}Os in K_2OsCl_6 at 4.2 K. In the oxides OsO_4 and K_2OsO_4, only lines corresponding to Ir entities iso-electronic with the parent compound were resolved, but a sizable fraction of the emission spectrum resides in a broad, largely featureless background. Radiation damage effects resulting from the $^{192}Os(n,\gamma)^{193}Os$ reaction were also very much in evidence. Jung and Triftshäuser (1968) report multiple line spectra for ^{125}Te following the K-capture decay of ^{125}I in $NaIO_3$ and $NaIO_4$. Their interpretation differs from that of Violet and Booth (1966) in that they invoke bond modifications rather than multiple charge states. On the other hand, the beta decay of ^{125}Sb in $NaSbO_3$ and Sb_2O_3 gave only single line ^{125}Te spectra. Similarly, Perlow and Perlow (1964) report the production of xenon compounds by the beta decay of iodine in iodine com-pounds. In every case the decay produced only a single chemical entity closely related to the iodine compounds. Herber and Stöckler (1964) report that the internal conversion of the 65.3 keV gamma ray following the decay of ^{119m}Sn in SnO_2 does not result in any chemical changes, but that in tetra-phenyl tin an additional component is observed. It was tentatively attributed to a Sn^+ ion in a molecule similar to $Sn(C_6H_5)_4$.

The study of these chemical consequences of nuclear events is as yet in an early state. Experiments on molecular crystals and hydrated salts show that fundamental chemical transformations, related to photolysis and radiolysis, can be studied by the ME. Experiments in ionic salts and in defect solids, e.g. the oxidic semiconductors, offer an approach to the study of defect states of a novel kind, whose origin and properties remain to be understood.

References

Ando, K. J., W. Kündig, G. Constabaris and R. H. Lindquist, 1967, J. Phys. Chem. Solids **28**, 2291.

Bearden, A. J., P. L. Mattern and T. R. Hart, 1964, Rev. Mod. Phys. **36**, 470.

Berger, W. G., 1969, Z. Physik **225**, 139.

Berger, W. G., J. Fink and F. E. Obenshain, 1967, Phys. Letters **A25**, 466.

Bhide, V. G. and G. K. Shenoy, 1966a, Phys. Rev. **143**, 309.

Bhide, V. G. and G. K. Shenoy, 1966b, Phys. Rev. **147**, 306.

Bondarevskiy, S. I. and P. P. Seregin, 1968, Fiz. Tverd. Tela **10**, 3454; Soviet Phys. Solid State, English transl. **10**, 2736.

Buchanan, D. N. E., 1970, J. Inorg. Nucl. Chem. **32**, 3531.

Carlson, T. A. and R. M. White, 1965, in: Proc. Symp. on Chemical effects associated with nuclear reactions and radioactive transformations, Vol. 1 (Intern. Atomic Energy Agency, Vienna).

Cavanagh, J. F., 1969, Phys. Stat. Sol. **36**, 657.

Chappert, J., R. B. Frankel and N. A. Blum, 1967, Phys. Letters **A25**, 149.

Chappert, J., R. B. Frankel, A. Misetich and N. A. Blum, 1969, Phys. Rev. **179**, 578.

Christiansen, J., E. Recknagel and G. Weyer, 1966, Phys. Letters **20**, 46.

Christiansen, J., P. Hindennach, U. Morfeld, E. Recknagel, D. Riegel and G. Weyer, 1967, Nucl. Phys. **A99**, 345.

Cohen, R. L., 1970, unpublished.

Czjzek, G., J. L. C. Ford, Jr., F. E. Obenshain and D. Seyboth, 1966, Phys. Letters **19**, 673.

Czjzek, G., J. L. C. Ford, Jr., J. C. Love, F. E. Obenshain and H. H. F. Wegener, 1967, Phys. Rev. Letters **18**, 529.

Dedericks, P. H., C. Lehmann and H. Wegener, 1965, Phys. Stat. Sol. **8**, 213.

De Waard, H. and S. A. Drentje, 1966, Phys. Letters **20**, 38.

De Waard, H., J. Heberle, P. J. Schurer, H. Hasper and F. W. J. Koks, 1968, in: E. Matthias and D. A. Shirley, eds., Hyperfine structure and nuclear radiation (North-Holland, Amsterdam) 329.

Dharmawardena, K. G. and G. M. Bancroft, 1968, J. Chem. Soc. A, 2655.

Drickamer, H. G., G. K. Lewis, Jr. and S. C. Fung, 1969, Science **163**, 885.

Dunlap, B. D., G. M. Kalvius, S. L. Ruby, M. B. Brodsky and D. Cohen, 1968, Phys. Rev. **171**, 316.

Eck, J. S., Y. K. Lee and J. C. Walker, 1967a, Phys. Rev. **163**, 1295.

Eck, J. S., Y. K. Lee, J. C. Walker and R. R. Stevens, 1967b, Phys. Rev. **156**, 246.

Erginsoy, C., G. H. Vineyard and A. Englert, 1964, Phys. Rev. **133**, A595.

Erginsoy, C., G. H. Vineyard and A. Shimizu, 1965, Phys. Rev. **139**, A118.

Fisher, B. and D. S. Tannhauser, 1966, J. Chem. Phys. **44**, 1663.

Fink, J. and P. Kienle, 1965, Phys. Letters **17**, 326.

Friedt, J. M. and J. P. Adloff, 1967, Compt. Rend. C **264**, 1356.

Friedt, J. M. and J. P. Adloff, 1969a, Compt. Rend. C **268**, 1342.

Friedt, J. M. and J. P. Adloff, 1969b, Inorg. Nucl. Chem. Letters **5**, 163.

Gibson, J. B., A. N. Goland, M. Milgram and G. H. Vineyard, 1960, Phys. Rev. **120**, 1229.

Goldberg, D. A., P. W. Keaton, Jr., Y. K. Lee, L. Madansky and J. C. Walker, 1965, Phys. Rev. Letters **15**, 418.

Goldberg, D. A., Y. K. Lee, E. T. Ritter, R. R. Stevens, Jr. and J. C. Walker, 1966, Phys. Letters **20**, 571.

Gütlich, P., S. O. Dar, B. W. Fitzsimmons and N. E. Erickson, 1968, Radiochim. Acta **10**, 147.

Hafemeister, D. W. and E. B. Shera, 1965, Phys. Rev. Letters **14**, 593.

Hannaford, P., 1968, Mössbauer studies of irradiation effects in solids, Ph. D. Thesis, Melbourne, Australia.

Hannaford, P. and J. W. G. Wignall, 1969, Phys. Stat. Sol. **35**, 809.

Hannaford, P., C. J. Howard and J. W. G. Wignall, 1965, Phys. Letters **19**, 267.

Hazony, Y. and R. H. Herber, 1969, J. Inorg. Nucl. Chem. **31**, 321.

Herber, R. H. and H. A. Stöckler, 1964, in: Chemical effects of nuclear transformations, Proc. Symp. on Chemical effects associated with nuclear reactions and radioactive transformations, Vol. 2 (Intern. Atomic Energy Agency, Vienna).

Ingalls, R. and G. de Pasquali, 1965, Phys. Letters **15**, 262.

Ingalls, R., C. J. Coston, G. de Pasquali and H. G. Drickamer, 1966, J. Chem. Phys. **44**, 1057.

Jacobs, C. G., Jr., N. Hershkowitz and J. B. Jeffries, 1969, Phys. Letters A**29**, 498.

Jagannathan, R. and H. B. Mathur, 1969a, Inorg. Nucl. Chem. Letters **5**, 89.

Jagannathan, R. and H. B. Mathur, 1969b, J. Inorg. Nucl. Chem. **31**, 3363.

Jagannathan, R., R. Thacker and H. B. Mathur, 1969, Indian J. Chem. **7**, 353.

Jung, P. and W. Triftshäuser, 1968, Phys. Rev. **175**, 512.

Klöckner, J., P. Rother, F. Wagner and U. Zahn, 1967, Angew. Chem. **6**, 1004.

Kröger, F. K. and H. J. Vink, 1956, Solid State Phys. **3**, 310.

Kündig, W., W. J. Ando, R. H. Lindquist and G. Constabaris, Czech. J. Phys. **17**, 467.

Lee, Y. K., P. W. Keaton, Jr., E. T. Ritter and J. C. Walker, 1965, Phys. Rev. Letters **14**, 957.

Mullen, J. G., 1965, Phys. Letters **15**, 15; Erratum: Ibid. **20** (1966) 320.

Mullen, J. G. and H. N. Ok, 1966, Phys. Rev. Letters **17**, 287.

Nath, A., R. D. Agarwal and P. K. Mathur, 1968a, Inorg. Nucl. Chem. Letters **4**, 161.

Nath, A., M. Harpold, M. P. Klein and W. Kündig, 1968b, Chem. Phys. Letters **2**, 471.

Nath, A., M. P. Klein, W. Kündig and D. Lichtenstein, 1970, Radiation Effects **2**, 211.

Nilsen, L., J. Lubbers, H. Postma, H. de Waard and S. A. Drentje, 1967, Phys. Letters **24B**, 144.

Ok, H. N. and J. G. Mullen, 1968a, Phys. Rev. **168**, 550, 563; Erratum: Ibid. **181** (1969) 986.

Ok, H. N. and J. G. Mullen, 1968b, Phys. Rev. Letters **21**, 823.

Perlow, G. J. and M. R. Perlow, 1964, in: Chemical effects of nuclear transformations, Proc. Symp. on Chemical effects associated with nuclear reactions and radioactive transformations, Vol. 2 (Intern. Atomic Energy Agency, Vienna).

Pollak, H., 1962, Phys. Stat. Sol. 2, 720.

Ritter, E. T., P. W. Keaton, Jr., Y. K. Lee, R. R. Stevens, Jr. and J. C. Walker, 1967, Phys. Rev. 154, 287.

Rother, P., F. Wagner and U. Zahn, 1969, Radiochim. Acta 11, 203.

Ruby, S. L. and R. E. Holland, 1965, Phys. Rev. Letters 14, 591.

Saito, N., H. Sano, T. Tominaga and F. Ambe, 1965, Bull. Chem. Soc. Japan 38, 681.

Sano, H. and F. Hashimoto, 1965, Bull. Chem. Soc. Japan 38, 1565.

Sano, H. and M. Kanno, 1969, J. Chem. Soc. D, 601.

Sauer, C., 1969, Z. Physik 222, 439.

Schroeer, D. and W. Triftshäuser, 1968, Phys. Rev. Letters 20, 1242.

Seyboth, D., F. E. Obenshain and G. Czjzek, 1965, Phys. Rev. Letters 14, 954.

Siegbahn, K., ed., 1965, Alpha-, beta- and gamma-ray spectroscopy, Vol. 1 (North-Holland, Amsterdam) 1523.

Siegwarth, J. D., 1967, Phys. Rev. 155, 285.

Sprouse, G. D., G. M. Kalvius and S. S. Hanna, 1967, Phys. Rev. Letters 18, 1041.

Stepanov, E. P. and A. Yu. Aleksandrov, 1967, Zh. Eksperim. i Teor. Fiz. Pis'ma 5, 101; Soviet Phys. JETP Letters, English Transl. 5, 83.

Stevens, R. R., J. S. Eck, E. T. Ritter, Y. K. Lee and J. C. Walker, 1967, Phys. Rev. 158, 1118.

Stone, J. A. and W. L. Pillinger, 1964, Phys. Rev. Letters 13, 200.

Temperley, A. A. and D. W. Pumplin, 1969, J. Inorg. Nucl. Chem. 31, 2711.

Triftshäuser, W. and P. P. Craig, 1966, Phys. Rev. Letters 16, 1161.

Triftshäuser, W. and P. P. Craig, 1967, Phys. Rev. 162, 274.

Triftshäuser, W. and D. Schroeer, 1969, Phys. Rev. 187, 491.

Trousdale, W. and P. P. Craig, 1967, Phys. Letters A27, 552.

Ullrich, J. F. and D. H. Vincent, 1969, J. Phys. Chem. Solids 30, 1189.

Varley, J. H. O., 1954a, Nature 174, 886.

Varley, J. H. O., 1954b, J. Nucl. Energy 1, 130.

Verwey, E. J. W., 1951, Semiconducting materials (Butterworths, London).

Vineyard, G. H., 1961, Discussions Faraday Soc. 31, 7.

Vineyard, G. H., 1963, J. Phys. Soc. Japan 18, Suppl. III, 144.

Violet, C. E. and R. Booth, 1966, Phys. Rev. 144, 225; Erratum: Ibid. 149, 414.

Wertheim, G. K. and D. N. E. Buchanan, 1969, J. Chem. Phys. 3, 87.

Wertheim, G. K. and H. J. Guggenheim, 1965, Chem. Phys. Letters 42, 3873.

Wertheim, G. K. and R. H. Herber, 1963, J. Chem. Phys. 38, 2106.

Wertheim, G. K., W. R. Kingston and R. H. Herber, 1962, J. Chem. Phys. 37, 687.

Wertheim, G. K., H. J. Guggenheim and D. N. E. Buchanan, 1968, Phys. Rev. Letters 20, 1158.

Wertheim, G. K., H. J. Guggenheim and D. N. E. Buchanan, 1969, J. Chem. Phys. 51, 1931.

Wiedemann, V. W., P. Kienle and F. Stanek, 1963, Z. Physik 15, 7.

Zimmerman, B. H., H. Jena, G. Ischenk, H. Kilian and D. Seyboth, 1968, Phys. Stat. Sol. 27, 639.

3 | IMPURITIES AND THEIR ASSOCIATION WITH DEFECTS IN NONMETALS

This chapter deals chiefly with the ME study of lattice defects and chemical impurities in nonmetals. There is an inevitable overlap with topics discussed in section 2.3 where the emphasis is on defect states *produced* by the radioactive decay of the parent of the Mössbauer isotope. In this chapter, the emphasis is on chemical impurities and their association with lattice defects in a variety of hosts. Most, but not all, of the experiments deal with ME absorbers containing the stable isotope. Experiments with sources are included only if ionization effects have been excluded, or if they serve to illuminate questions raised by experiments with absorbers.

3.1. Halides

The alkali and silver halides have been a favorite group of materials for the study of crystalline defects. Their simple structure, ease of preparation, stability and convenient optical properties have no doubt been factors in their selection. It is not surprising then that their defect properties were also investigated by the ME within a few years of its discovery. The most detailed ME investigations have been made in the AgCl system, and since these help to resolve some of the questions raised by the earlier alkali halide work we shall discuss them first.

The properties of dilute transition metal impurities in alkali and silver halides have been studied by ionic conductivity, electron paramagnetic resonance and optical techniques. The most detailed information is available for Mn^{2+} in alkali chlorides from the work of Watkins (1959), which gives information on the jump-time of the charge compensating vacancy that accompanies a divalent transition metal ion in these materials.

A number of investigators have carried out ME experiments on AgCl slabs prepared by diffusion-doping with ^{57}Co in a mixed argon and chlorine atmosphere. Surface contamination was generally removed by etching. This procedure introduces Co^{2+} ions with charge compensating Ag vacancies in (100) directions. The electron capture decay of $^{57}Co^{2+}$ apparently produces only $^{57}Fe^{2+}$, without any change in the lattice site or charge compensation.

The most clear-cut results in this system were obtained by Lindley and Debrunner (1966), whose data are reproduced in fig. 3.1. At low temperature, there is a doublet with an isomer shift characteristic of ionic Fe^{2+}. The quadrupole splitting (QS) is due to the destruction of cubic symmetry by the charge compensating vacancy. As the temperature is raised, the QS decreases slowly at first due to changes in the population of the crystal field levels and then rapidly as thermally activated vacancy motion sets in. At high temperature, a single line is obtained. These results and their interpretation are in essential agreement with the work of Hennig et al. (1966), see also table 3.1. The vacancy jump-time deduced from the ME measurements is in good agreement with that based on the dielectric loss. Absorber experiments with ^{57}Fe in AgCl, carried out by Hennig (1968), offer additional evidence that the identification of the Fe^{2+}–Ag vacancy pair is correct.

Lindley and Debrunner (1966) also found that room temperature aging introduces two new spectra, one corresponding to divalent iron with larger QS, the other to trivalent iron with negligible QS. The divalent spectrum resembles that found in oxygen-doped crystals, suggesting that it is due to association of Co^{2+} with O^{2-}. Similar changes on heating in helium saturated with water vapor have been reported by Hennig et al. (1968).

Generally similar results have been obtained in AgCl by Murin et al. (1967) but their interpretation differs with regard to the identification of ME lines with defect configurations. For example, the spectrum previously identified as due to a substitutional Co^{2+} with Ag vacancy charge compensation is here attributed to interstitial Co^{2+}. There appears to be no independent evidence, however, for the existence of such high relative concentrations of divalent transition metal impurities in interstitial sites in AgCl. By diffusion in air, or by using highly diluted $CoCl_2$ solution, they produce a spectrum suggestive of trivalent iron, on the basis of both the IS and QS. However, they identify it as due to covalent substitutional Co^{2+} with Ag vacancy charge compensation. It is more likely that it is due to oxygen charge compensation. At high cobalt concentration they obtained a precipitated $CoCl_2$ phase similar to that reported by Mullen (1963) in NaCl.

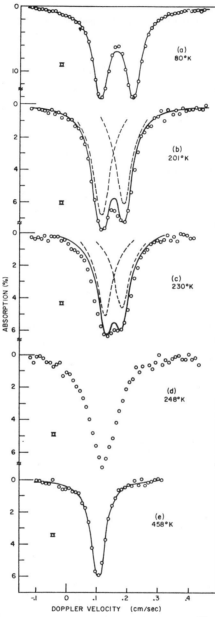

Fig. 3.1. Mössbauer spectra for a freshly prepared sample of ^{57}Co in AgCl at five temperatures. The absorber was enriched stainless steel. The solid lines in (a), (b) and (c) are two line spectra whose components are Lorentzian and have widths of 0.06 cm/s. The solid line in (e) is a Lorentzian of width 0.059 cm/s. (From Lindley and Debrunner, 1966.)

TABLE 3.1

Material	Refs.	Temperature (K)	IS_{Fe} (cm/s)	QS (cm/s)		Remarks
AgCl	[1]	77	0.123	0.106		(data read from
		200	0.113	0.073	(a)	
		300	0.105	≈ 0		graph)
		450	0.097	0		
AgCl	[2]	77	0.132	0.107		—
		195	0.125	0.074	(a)	
		297	0.115	< 0.03		
AgCl	[3]	80	0.132	0.105		—
		200	0.126	0.090	(b)	
		295	0.113	< 0.03		
AgCl	[1]	80	0.101	0.24		oxygen doped
AgCl	[1]	80	0.074	—		precipitated phase
AgCl	[3]	80	0.056	0.04	(a)	—
KCl	[4]	295	0.048	≈ 0.04		quenched
		295	0.033	0.068	(c)	slow cooled
NaCl	[5]	80	0.041	0.055	(d)	slow cooled
		295	0.034	0.051	(d)	slow cooled
NaCl	[6]	300	0.110	0.210		converts to Fe^{3+} in wet He

(a) identified by authors as substitutional cobalt with cation vacancy charge compensation

(b) identified as due to interstitial cobalt

(c) also shows line at 0.2 cm/s

(d) also shows line at 0.2 cm/s identified by author as Fe^+

[1] Lindley and Debrunner (1966)
[2] Hennig et al. (1966)
[3] Murin et al. (1967)
[4] De Coster and Amelinckx (1962)
[5] Mullen (1963)
[6] Hennig et al. (1968)

Thus, except for questions of identification, which may be resolved in the way indicated, a clear picture of the behavior of Co^{2+} and Fe^{2+} in AgCl emerges from the ME experiments. Substitutional Co^{2+} with Ag vacancy charge compensation dominates in samples prepared in the absence of

oxygen and water vapor. The effect of the motion of the vacancy is understood quantitatively. The presence of oxygen leads either to substitutional
Co^{2+} with O^{2-} charge compensation, or to trivalent iron.

With this information, it is instructive to consider the ME studies of
KCl by de Coster and Amelinckx (1962), of NaCl by Mullen (1963) and
Hennig et al. (1968), and of NaF by Wertheim and Guggenheim (1965).
The spectra obtained (table 3.1) generally fall into two categories, a barely
split doublet with an isomer shift of 0.04 cm/s indicative of trivalent iron
and a well split doublet (QS $\approx$ 0.2 cm/s) with an IS of 0.11 cm/s indicative
of divalent iron.

The identification of the peak at 0.2 cm/s in KCl and NaCl no longer
poses a particular problem. Hennig et al. (1968) have obtained a divalent
iron spectrum with lines at 0.005 and 0.215 cm/s in NaCl. The latter is,
within the error, identical with the line reported in KCl and the so-called
B-spectrum of NaCl. Other facts also support this identification. Hennig
and coworkers report that the divalent spectrum is converted into the trivalent one by annealing in wet helium or air, indicating that the latter is due
to oxygen charge compensation. (The divalent spectrum presumably corresponds to substitutional divalent cobalt with cation vacancy charge compensation.) With this in mind, the increased intensity of the line at 0.2 in
KCl and NaCl due to quenching follows naturally. A rapid quench prevents
the pairing of Co^{2+} with O^{2-}. The suggestion that the line at 0.2 is due to
Co^+ is not in accord with other evidence which shows that this valence state
is not obtained by quenching. It has, however, been obtained in low concentration by ionizing radiation.

In NaF containing ^{57}Co introduced by diffusion in HF, only divalent
iron was observed. Quadrupole splitting was attributed to a charge compensating Na vacancy. It was also suggested that a variety of divalent spectra
which were barely resolved at 78 K could be attributed to the various neighboring lattice sites for the vacancy. Annealing in air produced trivalent iron
with a spectrum showing broadened magnetic hfs. It was suggested that the
splitting is due to isolated iron ions with long spin relaxation time, but it
is also possible that extended annealing in air leads to the precipitation of
oxides on dislocations or even on the surface. This aspect of the experiment
deserves further attention.

In summary, ME measurements on ^{57}Fe produced by the decay of
^{57}Co in alkali and silver halides show that suitably prepared samples contain the transition metal ion in a substitutional site with a near-neighbor
cation vacancy. The effect of the vacancy and its motion on the QS and

its temperature dependence supports this interpretation. Annealing in the presence of water vapor introduces another form of charge compensation, presumably oxygen, and favors the formation of trivalent iron. Other valences of iron remain to be identified in a convincing manner.

3.2. Oxides

Iron as a chemical impurity has also been studied by the ME in a number of oxides. Simplest are the sodium chloride structure MO compounds into which Fe^{2+} can be introduced as a substutitional impurity on a cubic site. Of equal interest are compounds with corundum, spinel, rutile or perovskite structures.

The most definitive study has been made in MgO by Leider and Pipkorn (1968). They show that both Fe^{2+} and Fe^{3+} are produced by diffusion in air, in agreement with earlier work (Wertheim and Buchanan, 1962), and that reduction by Mg vapor or by a $CO + CO_2$ atmosphere converts the iron almost completely to the divalent state. The behavior of the Fe^{2+} has been studied in considerable detail. At high temperature, a single line, appropriate for an undistorted cubic environment, is found. Below 20 K, the line broadens and eventually splits into a quadrupolar doublet. The splitting is attributed to the lifting of the degeneracy of the Γ_{5g} ground state triplet by random strain, and the slowing down of the relaxation between the levels of the triplet at low temperature.

FeO is a defect solid in which excess oxygen is always present. The ME spectrum of the iron exhibits quadrupole splitting which has been attributed to the presence of metal ion vacancies (Shirane et al., 1962). These may produce a field gradient through a mechanism similar to that in MgO. The magnitude of the splitting is the same in the two cases.

More recent investigations have shown that the spectrum of FeO can be decomposed into two overlapping quadrupole-split patterns whose relative intensity is a function of composition (Elias and Linnett, 1969; Johnson, 1969). The latter concludes that the discrete nature of the spectra implies the existence of two ordered defect structures which he associates with n- and p-type FeO. Elias and Linnett, on the other hand, propose that the two spectra rise from ions in two distinct lattice sites, octahedral and tetrahedral. (The tetrahedral sites are similar to those of Fe_3O_4.) It is shown that the changes in isomer shifts and line intensity which accompany changes in composition can be understood if electron charge hopping among octahedral site ions, and between tetrahedral site ions and their immediate neighbors is invoked. The interpretations offered by Elias and Linnett and by Johnson appear difficult to reconcile unless it should turn out that the region around a tetrahedral site is in fact n-type. But even then a one-to-one correspondence between the two models is not achieved.

Iron is readily introduced into Al_2O_3 by diffusion in air. Uniform low concentrations are more readily achieved by growth from a flux. In material of the latter kind, the ME hyperfine structure of dilute Fe^{3+} in the three crystal field states was first demonstrated by Wertheim and Remeika (1964, 1965). In a detailed study of the effects of heat treatment, Bhide and Date (1968) investigated the reduction and clustering of the iron. They found that vacuum firing produces Fe^{2+} with properties identical to those obtained in ^{57}Co doped Al_2O_3. Firing in hydrogen produced metallic iron in ferromagnetic clusters. Subsequent air firing was found first to convert the iron to α-Fe_2O_3 and then to produce dissolved Fe^{3+}.

The ME of ^{57}Fe as a dilute impurity has also been studied in the compounds V_2O_3 and VO_2 which are of interest because of the metal-insulator transitions near 160 K and 340 K, respectively. In V_2O_3, Shinjo and Kosuge (1966) report antiferromagnetism below the transition. In VO_2, both Kosuge (1967) and Wertheim et al. (1967) report two types of trivalent iron which exhibit discontinuous behavior in IS and QS at the transition. The two Fe^{3+} spectra differ largely in the magnitude of the QS, but both types of iron are present in equal amounts. This suggests that they arise from association with the charge compensating oxygen vacancy by one of the two iron atoms corresponding to each vacancy. The QS is quite small in the metallic state.

Evidence for vacancy association has also been obtained in the perovskites $BaTiO_3$ and $SrTiO_3$. These were studied as sources containing ^{57}Co introduced by diffusion. The transition metals cobalt and iron are known to enter the materials substitutionally for Ti^{4+}. The isostructural compound $SrFeO_3$, in which iron is also four-valent, has been studied by the ME by Gallagher et al. (1964). Their data provide a valuable reference point which aids in the identification of the valence states by their IS and QS. The general tendency in the titanates is toward the stabilization of Fe^{3+} by oxygen vacancy charge compensation. However, various states can be produced, depending on the heat treatment which is employed.

Single crystals of $BaTiO_3$ annealed for two hours in air and then quenched show no effects of vacancy association (Bhide and Multani, 1965). The iron is clearly trivalent, and the small QS which vanishes at the ferroelectric Curie point is attributed to the distortion of the cubic perovskite in the ferroelectric state. In slow-cooled samples, on the other hand, two overlapping spectra are detected, one like that in the quenched state, and the other with much larger QS. Both have IS's which show them to be trivalent, but the QS of the second type does not vanish above the Curie point. This is good evidence that it is due to a nearby lattice defect, presumably an oxygen vacancy. The spectrum of $BaTiO_3$ annealed in hydrogen and then slow-cooled exhibits additional lines indicative of divalent iron.

In single crystal $SrTiO_3$ into which ^{57}Co had been introduced by an anneal in hydrogen followed by slow cooling, one or two ME absorption lines were found at room temperature, depending on the details of the heat treatment (Bhide and Bhasin, 1967). (This compound retains the cubic perovskite structure down to 110 K, where it becomes tetragonal.) It is shown that the two lines represent two different states of the iron atom and that neither one has resolved QS. Increased duration of hydrogen firing shifts the intensity toward the line with characteristic Fe^{3+} isomer shift. The other has a much smaller shift. The authors argue that this line corresponds to trivalent iron in a low-spin state, in part on the basis that it has an IS similar to that of iron in the ferricyanide complex. Other support is drawn from the fact that trivalent cobalt is diamagnetic in $SrTiO_3$.

It is difficult to rule out the alternative that the line with small IS in fact corresponds to Fe^{4+}. While one may wish to reject this notion simply on the basis that an anneal in hydrogen will never allow such a valence state, it must be remembered that it is the cobalt parent which is introduced during the anneal. The absence of quadrupole splitting indicates that the charge compensating oxygen vacancies are not associated with the transition metal ion, i.e. that the iron atom finds itself in a normal lattice site of the perovskite. Since the compound $SrFe^{4+}O_3$ exists, it would not be surprising to find that the iron assumes the 4+ valence, especially since the ionic radius of Fe^{3+} (0.67 Å) is slightly greater than that of Ti^{4+} (0.64 Å). Comparison with the result of Gallagher et al. (1964) (see table 3.2) shows close correspondence with the properties of Fe^{4+} in $SrFeO_3$. One could distinguish between these two alternatives by studying the splitting of this line in a high magnetic field at low temperature. The saturation value of the internal field for Fe^{4+} should be ≈ 330 kOe as in $SrFeO_3$ while that of low spin trivalent iron should be smaller.

In $SrTiO_3$ quenched from high temperature, evidence for Fe^{2+} and Fe^{3+} with QS is obtained. This is in accord with expectations since this process results in partial reduction, i.e. introduction of oxygen vacancies.

Some experiments with ^{57}Fe doped $SrTiO_3$ have been reported by

TABLE 3.2

Compound and treatment		Refs.	$1S_{Fe}$ [a] (cm/s)	QS [a] (cm/s)
$BaTiO_3$ air anneal, quench		[1]	0.041_2	0.046
$BaTiO_3$ air anneal, slow cool		[2]	0.040_5	0.042
			0.0025	0.112
$SrTiO_3$ hydrogen anneal, slow cool		[3]	0.036_4	0
			0.009 [b]	0
$SrFeO_3$	Fe^{4+}	[4]	0.0055	0
$SrFeO_{2.86}$	Fe^{4+}	[4]	0.0022	0
	Fe^{3+}	[4]	0.0485	—

[a] at room temperature
[b] identified as low-spin Fe^{3+}

[1] Bhide and Multani (1965)
[2] Bhide and Multani (1966)
[3] Bhide and Bhasin (1967)
[4] Gallagher et al. (1964)

Bhide et al. (1967) and Bhide and Bhasin (1967). The concentration of the transition metal is, of course, many orders of magnitude higher, 1.5 to 3 mole %. In air-fired material, the iron is present as Fe^{3+} with anion vacancy charge compensation. Vacuum firing leads to the production of Fe^{2+} with large QS. Hydrogen firing leads to the precipitation of initially colloidal metallic iron which then agglomerates to form ferromagnetic particles with characteristic hfs.

The ferroelectric phase transition in $BaTiO_4$ has also been studied using ^{119}Sn substituted for titanium (Chekin et al., 1965). The sample used contained 1 at.% Sn, a concentration sufficiently low so that the ferroelectric properties are substantially unaffected. The recoil-free fraction exhibited a sharp decrease as the Curie point was approached from the paraelectric region. This is in agreement with theoretical predictions based on the anomalous optical branch which gives rise to the phase transition. Similar results at higher tin concentration have been reported by Belov and Zheludev (1967).

3.3. Sulphides

A comparison of luminescent and ME properties of iron in ZnS was made by Luchner and Dietl (1963, 1964). They diffused ZnS coated with isotopically enriched ferrous sulphide in air or nitrogen at temperatures between 400 and 1100 °C and found successive transformations of the lattice site of the iron. It was found possible to make a correspondence between particular ME absorption spectra and certain luminescence centers produced by iron.

The authors conclude on the basis of the IS that all of the iron found in the ME spectra is trivalent. The most prominent narrow line, labeled III, has an IS of ≈ 0.07 cm/s at room temperature which is indeed much closer to the ionic trivalent iron value (0.05 cm/s) than to the divalent one (0.14 cm/s). However, it is larger than the limiting value of Fe^{3+} and therefore suspect. In the light of current knowledge that the IS of Fe^{2+} in tetrahedral sulphur coordination in $FeCr_2S_4$ is 0.060 cm/s and that it is only slightly larger (0.080 cm/s) in octahedral coordination in $FeIn_2S_4$ (Eibschütz et al., 1967), one tends to assign this line to Fe^{2+}. This suggestion is in line with finding that this type of iron is formed only until the divalent iron on the surface is exhausted; it is not associated with luminescence properties. See also the discussion of $ZnS : Co^{57}$ below.

The luminescence properties were associated with two ME spectra which were clearly due to Fe^{3+}. The one with vanishing quadrupole splitting, labeled II, corresponds to the red emission, and the one labeled IV to a quenching center. The quadrupole splitting of the latter was ascribed to association with a Zn vacancy.

Although the behavior of the ZnS : Fe system is quite complex, the results already obtained suggest that extension of this work should be of great interest. Attempts to change the valence, by heat treatment in a reducing atmosphere or in sulphur vapor, might shed some further light on the identification of these centers. Attempts to study the ME of optically excited states also appear feasible in this or related systems.

The emission spectrum of ^{57}Co doped ZnS was studied by Belozerskiy et al. (1966), who found only a single line which appears to correspond closely to line III in the absorber experiment discussed above. The authors identify this line with substitutional Fe^{2+} produced by the decay of Co^{2+} which is in agreement with the reinterpretation of the $ZnS : {}^{57}Fe$ data suggested above.

3.4. Semiconductors

The electronic properties of dilute chemical impurities in semiconductors are well established. Many of them have discrete electronic levels in the forbidden gap and therefore exist in different charge states in n- and p-type material. Unlike the normal donors or acceptors which bind electrons in hydrogenic orbits of large radius, the 'deep' donors and acceptors have compact wavefunctions. As a result it might be argued that the localization of an electron on such a donor or acceptor should result in a measurable change in the ME isomer shift.

To date, experiments have been carried out with ^{57}Fe and ^{119}Sn in a variety of semiconductors. The solubility of iron in most semiconductors is, however, too small to allow a direct ME absorption experiment. Instead, it has generally been necessary to prepare ^{57}Co doped semiconductor sources. This should not vitiate the experiment in those cases where it is known that iron and cobalt enter the semiconductor lattice in the same way. Except for carrier trapping effects which should be important mainly at low temperature, electronic relaxation should proceed rapidly compared to the 10^{-7} s half-life of the first excited state of ^{57}Fe which emits the gamma ray. Minority carrier trapping would be particularly serious in an n-type semiconductor with a double acceptor, e.g. iron or cobalt in germanium. Holes created by the Auger process would be annihilated quite slowly by electrons from the conduction band because the final stage of relaxation requires capture of an electron by an already negatively charged defect. The import of trapping to the interpretation of the Mössbauer experiments remains to be examined in detail.

Experiments of this type in the elemental semiconductors Si and Ge have been reported by Norem and Wertheim (1962) and in Ge by Belozerskiy et al. (1966). The spectra in n- and p-type silicon were relatively broad (0.06 cm/s) but no difference greater than 0.0003 cm/s was detected in the IS. A subsidiary line in Si was identified with substitutional iron. Its IS is similar to that of substitutional iron in the III–V semiconductors (table 3.3). This line was enhanced by the presence of copper which is known to generate vacancies in silicon. Copper is substitutional at high temperature but becomes interstitial at low temperature, leaving vacancies after a quench. These vacancies are mobile at room temperature and apparently annihilate with interstitial iron making it substitutional. In both n- and p-type germanium, the spectrum was a doublet with QS = 0.042 cm/s and IS = 0.036 cm/s.

A number of the III–V semiconductors have also been examined, e.g.,

TABLE 3.3

Semiconductor	Refs.	IS_{ss} [a] (cm/s)		QS [a] (cm/s)	Γ [a] (cm/s)
Si	[1]	Int.	+0.004	0	≈ 0.06
		Sub.	0.059	< 0.03	≈ 0.06
Ge	[1]		0.041	0.042	0.06
Ge	[2]		0.040	[b]	0.10
InSb	[3]		0.044	0	0.051
InP	[4]		0.036–0.37	0	0.059–0.60
GaAs	[4]		0.036–0.37	0	0.059–0.60
GaAs	[5]	I	0.045	0.035	≈ 0.04
		II	0.037	0.125	0.04
InAs	[6]		0.051	0.045	—
InAs	[7]		0.047	0.046	0.053
GaP	[7]		0.031	0	0.060

[a] all values at room temperature
[b] not resolved

[1] Norem and Wertheim (1962)
[2] Belozerskiy et al. (1966)
[3] Belozerskiy et al. (1965a)
[4] Belozerskiy et al. (1965b)
[5] Albanese et al. (1967)
[6] Bemski and Fernandes (1963)
[7] Basetskiy et al. (1968)

InAs by Bemski and Fernandes (1963) and Basetskiy et al. (1968), InSb by Belozerskiy et al. (1965a), InP and GaAs by Belozerskiy et al. (1965b), GaAs by Albanese et al. (1967), and GaP by Basetskiy et al. (1968). These workers all agree in finding no difference between n- and p-type semiconductors. There are, however, some significant differences among these materials, see table 3.3. Most striking is the question of the QS in the III–V compounds. Albanese et al. have shown that the broad lines universally reported in fact consist of two sets of quadrupolar doublets. They accomplished this by using an absorber ($FeGe_2$) which has a better linewidth than the stainless steel used in the other investigations, and by examining the emission spectrum as a function of the depth of diffusion. However, even in this experiment no difference was found in the IS between n- and p-type for either of the two quadrupolar doublets.

The value of the IS, which is close to that expected for a covalently bonded Fe^{3+} ion, suggests that iron enters the III–V compounds in the $3d^5$ configuration substitutionally for the group III atom. This is reasonable from the point of view of the electronic configurations (Ga: $3d^{10}4s^24p^1$; Fe: $3d^64s^2$; As: $3d^{10}4s^24p^3$) since iron would then contribute three electrons toward the formation of sp^3 tetrahedral bonds. In GaAs, a second type of iron with similar IS but larger QS is thought to be due to association of Co with divacancies. These vacancies would also account for the more rapid diffusion of the cobalt which produces this type of iron.

So far, a unique explanation of the lack of difference in IS for iron in n- and p-type semiconductor hosts has not been given. We consider only those hosts where iron has one or more levels in the forbidden gap and accept the suggestion that the iron seen in the ME is substitutional, i.e. like the iron seen in electrical measurements. We then conclude either that the additional electron bound by the iron acceptor does not produce any change in the electronic charge density at the nucleus, perhaps because of balanced s and d character or because its wavefunction extends much further from the nucleus than the atomic wave functions, or that trapping prevents the attainment of the equilibrium charge state.

Isomer shifts of ^{119}Sn in various semiconductors have been measured by Chekin et al. (1967) but a detailed interpretation has not been given.

References

Albanese, G., G. Fabri, C. Lamborizio, M. Musci and I. Ortalli, 1967, Nuovo Cimento B50, 149.

Basetskiy, V. Ya., B. N. Veits, V. Ya. Grigalis, Yu. D. Lisin and I. M. Taksar, 1968, Fiz. Tverd. Tela 10, 2852; Soviet Phys. Solid State, English Transl. 10 (1969) 2252.

Belov, V. F., and I. Z. Zheludev, 1967, Zh. Eksperim. i Teor. Fiz. Pis'ma 6, 843; Soviet Phys. JETP Letters, English Transl. 6 (1967) 287.

Belozerskiy, G. N., I. A. Gusev, A. N. Murin and Yu. A. Nemilov, 1965a, Fiz. Tverd. Tela 7, 1254; Soviet Phys. Solid State, English Transl. 7 (1965) 1012.

Belozerskiy, G. N., Yu. A. Nemilov, S. B. Tomilov and A. V. Shvedchikov, 1965b, Fiz. Tverd. Tela 7, 3607; Soviet Phys. Solid State, English Transl. 7 (1966) 2908.

Belozerskiy, G. N., Yu. A. Nemilov, S. B. Tomilov and A. V. Shvedchikov, 1966, Fiz. Tverd. Tela 8, 604; Soviet Phys. Solid State, English Transl. 8 (1966) 485.

Bemski, G. and J. C. Fernandes, 1963, Phys. Letters 6, 10.

Bhide, V. G. and H. C. Bhasin, 1967, Phys. Rev. 159, 586.

Bhide, V. G. and S. K. Date, 1968, Phys. Rev. **172**, 345.

Bhide, V. G. and M. S. Multani, 1965, Phys. Rev. **139**, A1983.

Bhide, V. G. and M. S. Multani, 1966, Phys. Rev. **149**, 289.

Bhide, V. G., H. C. Bhasin and G. K. Shenoy, 1967, Phys. Letters **A24**, 109.

Chekin, V. V., V. P. Romanov, B. I. Verkin and V. A. Bokov, 1965, Zh. Experim. i Teor. Fiz. Pis'ma **2**, 186; Soviet Phys. JEPT Letters, English Transl. **2** (1965) 117.

Chekin, V. V., A. P. Vinnikov and O. P. Balkashin, 1967, Fiz. Tverd. Tela **9**, 2992; Soviet Phys. Solid State, English Transl. **9** (1968) 2354.

De Coster, M. and S. Amelinckx, 1962, Phys. Letters **1**, 245.

Eibschütz, M., E. Hermon and S. Shtrikman, 1967, Solid State Commun. **5**, 529.

Elias, D. J. and J. W. Linnett, 1969, Trans. Faraday Soc. **65**, 2673.

Gallagher, P. K., J. B. Mac Chesney and D. N. E. Buchanan, 1964, J. Chem. Phys. **41**, 2429.

Hennig, K., 1968, Phys. Stat. Sol. **27**, K115.

Hennig, K., W. Meisel and H. Schnorr, 1966, Phys. Stat. Sol. **13**, K9; Ibid. **15** (1966) 199.

Hennig, K., K. Yung and B. S. Skorchev, 1968, Phys. Stat. Sol. **27**, K161.

Johnson, D. P., 1969, Solid State Commun. **7**, 1785.

Kosuge, K., 1967, J. Phys. Soc. Japan **22**, 551.

Leider, H. R. and D. N. Pipkorn, 1968, Phys. Rev. **165**, 494.

Lindley, D. H. and P. G. Debrunner, 1966, Phys. Rev. **146**, 199.

Luchner, K. and J. Dietl, 1963, Z. Physik **176**, 261.

Luchner, K. and J. Dietl, 1964, Acta Phys. Polon. **26**, 697.

Mullen, J. G., 1963, Phys. Rev. **131**, 1415.

Murin, A. N., B. G. Lure and P. P. Seregin, 1967a, Fiz. Tverd. Tela **9**, 2428; Soviet Phys. Solid State, English Transl. **9** (1968) 1901.

Murin, A. N., B. G. Lure and P. P. Seregin, 1967b, Fiz. Tverd. Tela **9**, 1424; Soviet Phys. Solid State, English Transl. **9** (1967) 1110.

Murin, A. N., B. G. Lure, P. P. Seregin and N. K. Cherezov, 1967c, Fiz. Tverd. Tela **8**, 3291; Soviet Phys. Solid State, English Transl. **8** (1967) 2632.

Norem, P. C. and G. K. Wertheim, 1962, J. Phys. Chem. Solids **23**, 1111.

Shinjo, T. and K. Kosuge, 1966, J. Phys. Soc. Japan **21**, 2622.

Shirane, G., D. E. Cox and S. L. Ruby, 1962, Phys. Rev. **125**, 1158.

Watkins, G. D., 1959, Phys. Rev. **113**, 79, 91.

Wertheim, G. K. and D. N. E. Buchanan, 1962, in: D. M. J. Compton and A. H. Schoen, eds., The Mössbauer effect (Wiley, New York) 130-140.

Wertheim, G. K. and H. J. Guggenheim, 1965, J. Chem. Phys. **42**, 3873.

Wertheim, G. K. and J. P. Remeika, 1964, Phys. Letters **10**, 14.

Wertheim, G. K. and J. P. Remeika, 1965, in: L. van Gerven, ed., Nuclear magnetic resonance and relaxation in solids, Proc. Coll. Ampère XIII (North-Holland, Amsterdam) 147-161.

Wertheim, G. K., D. N. E. Buchanan and H. J. Guggenheim, 1967, Bull. Am. Phys. Soc. **12**, 23.

4 | CHEMICAL IMPURITIES IN METALS

There are two distinct aspects to the study of dilute chemical impurities in metals. The first is the perturbation of the host lattice by a defect and the second the properties of the impurity atom itself. A great deal of effort has gone into the ME study of both parts of this problem. We will first consider the effects of substitutional and interstitial chemical defects in metals, especially iron, where the majority of such studies has been made.

4.1. Effects on the host lattice

4.1.1. SUBSTITUTIONAL IMPURITIES IN IRON

Early investigations in an iron-based alloy system showed quite clearly that the ME can be used to discern the effects of dilute solute impurity elements on the host lattice (Johnson et al., 1963). Satellite structure on the normal hfs of metallic iron was first reported in the Fe–Si systems (Stearns, 1963), and subsequently for a wide range of other solutes (Wertheim et al., 1964; Stearns and Wilson, 1964; Stearns, 1966; Cranshaw et al., 1966; Bernas and Campbell, 1966; Marcus and Schwartz, 1967).

Data analysis has been based on a model in which impurity atoms in the various coordination shells of a host lattice atom are assumed to produce additive changes in its hfs effective field H_{eff}. Although analyses have been done considering only one or as many as six shells, most of the data have been analyzed by considering one or two shells plus a concentration dependent term. This is satisfactory since four- and six-shell calculations show that the effects of impurity atoms in the outer shells is very much smaller than that of the near and next-near neighbor shells. Moreover, nmr experiments which offer much greater resolution have failed to confirm the existence of outer shell satellites deduced from ME experiments (Rubinstein et al., 1966). Quadrupolar effects have generally been too small to resolve. The assumption of linearity and superposition implicit in these analyses have not been tested in detail.

The relationship between H_{eff} and the magnetic moment of the iron atom μ has been clarified by comparison with neutron diffraction data due to Collins and Low (1964, 1965) and Low and Collins (1963). It is found that the perturbation in μ is much smaller than that in H_{eff}, indicating that the major part of the effect in the Mössbauer hfs arises from changes in the conduction electron polarization (Stearns and Wilson, 1964; Campbell, 1966; Stearns, 1966). Consequently, the value of H_{eff} obtained from Mössbauer spectra does not give direct indication of the moment on the atom. Nevertheless it remains a useful parameter which has been used in the study of short-range order, order–disorder transformation and precipitation. These are discussed in section 4.3.

4.1.2. Interstitial impurities in iron

The effect of interstitial impurities on the iron host lattice is somewhat different. We consider the Fe–C system in which a number of ME studies have recently been reported.

In this system the solubility of C in fcc iron extends to 2.1 wt. % (9 at. %) at 1150 °C (Pearson, 1958). Upon rapid quenching, this phase, called austenite, transforms by a diffusionless process to a nonequilibrium bct phase called martensite. Annealing below 727 °C leads to the precipitation of carbides (e.g., Fe_3C, orthorhombic cementite) in a bcc α-iron matrix called ferrite. The carbides are themselves metastable. The true Fe–C equilibrium phase diagram contains only graphite and iron solid solution phases, but this condition is not approached in the experiments to be considered below.

The characteristic ME spectra of these phases are well established. Austenite is nonmagnetic and has an unsplit component with an IS slightly smaller than that of pure iron and a quadrupole split component corresponding to iron atoms with carbon neighbors. The spectrum of ferrite is indistinguishable from that of high-purity iron since the solubility of C in α-iron is very small. Martensite has a spectrum superficially similar to that of pure iron but with broadened lines and some satellite structure. Cementite is also magnetically ordered at room temperature but its hfs corresponds to an effective magnetic field $\approx \frac{2}{3}$ that of pure iron. It is therefore clear that the transformations in the Fe–C system can be studied with facility by the ME.

The spectrum of retained austenite in quenched high carbon steel has been reported to consist of a singlet with an IS similar to that of austenitic stainless steels and a quadrupole split doublet (Gielen and Kaplow, 1967; Christ and Giles, 1968). The doublet could be related to the early stages of carbide precipitation but most probably represents iron atoms with C neighbors. Lewis and Flinn (1968) have shown that the QS disappears at high temperature where the jump frequency of the carbon becomes so high that all iron atoms see, in effect, the same time-average environment (fig. 4.1). This result argues against the identification of the doublet with precipitated carbides.

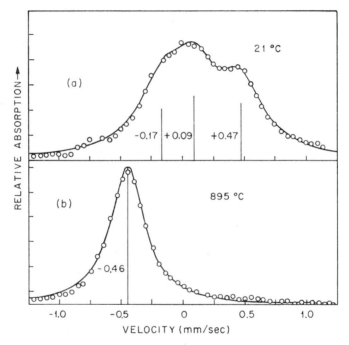

Fig. 4.1. Mössbauer spectra of austenite. (a). Retained austenite at room temperature showing the single line due to iron atoms without carbon neighbors and the doublet due to those with carbon neighbors. (b). Austenite at 895 °C showing that all iron atoms have a uniform environment. The shift of the spectrum to lower energy is due to the second-order Doppler effect. (From Lewis and Flinn, 1968.)

There is satisfactory agreement among the spectra reported for marten-site quenched to room temperature, including those of Zemcik (1967), Gielen and Kaplow (1967), Ino et al. (1967) and Moriya et al. (1968). The latter have analyzed the spectrum of martensite in detail on the assumption that neighbors beyond the fourth are unaffected by the presence of carbon and have spectra like those of α-iron. The validity of this assumption should be examined with care since the tetragonal lattice distortion and electronic effects may not be negligible. In the case of substitutional impurities there is good evidence for long range effects which would invalidate some of the detail obtained by this method of analysis. A more fundamental question is raised by Genin and Flinn (1966) who showed that there are changes in the ME spectra during the early stages of room temperature aging of quenched martensite. They suggest that this is due to the clustering of carbon atoms. In some other investigations, it was assumed that martensite is stable at room temperature and has statistically random distribution of carbon atoms. Thus a weak six-line spectrum which shows significant quadrupole splitting is ascribed to C–C clusters by Genin and Flinn and to iron atoms near-neighbor to a single C atom by most other investigators. Ron et al. (1967), who also studied martensite cooled rapidly to liquid nitrogen temperature, ascribe this line to iron atoms with a single C atom along the c-axis. However, their analysis differs in other aspects from that of Moriya et al. (1968). This brief survey should suffice to indicate that there are significant unresolved problems in the details of the interpretation of the ME spectrum of martensite. Ultimately the interpretation hinges on the question of whether the carbon remains randomly distributed at room temperature, and whether the spectrum can be decomposed unambiguously into contributions from the various coordination shells around a carbon.

The ME properties of cementite were established by Shinjo et al. (1964), Ron et al. (1966), Bernas et al. (1967), and Ron et al. (1968). Bernas et al. used carbides prepared by reaction of iron oxide with a $CO+H_2$ mixture, the others examined carbides isolated from their iron matrix by electrolytic or chemical etching. It is known that cementite itself has a range of composition and there are other carbide phases, e.g. hexagonal ε-Fe_xC and χ-carbide. Ron et al. (1968) examined carbides formed during annealing and found that the ME properties of carbide precipitates depend sensitively on the annealing temperature. The precipitate obtained by low temperature annealing appears superparamagnetic. Those obtained at higher temperature have a systematically varying H_{eff} and T_c. Above T_c, the spectrum consists of a doublet suggestive of QS but the spectrum in the magnetically ordered state gives no evidence for quadrupole interaction. The apparent absence of QS in the ordered state could result simply from having the magnetization at an angle near 54° 44′ with respect to the symmetry axis of the EFG tensor.

This would allow the doublet structure observed in the initial stages of precipitation to be identified with superparamagnetic carbide.

The transformations during tempering of iron–carbon martensite have also been studied by Ino et al. (1968). They identify (1) ε-carbide formed from martensite after a one-hour anneal at 140 °C, (2) the conversion of retained martensite and partial conversion of ε-carbide to χ-carbide after one hour at 220 °C, (3) the complete conversion of ε-carbide to χ-carbide after one hour at 340 °C and then (4) to cementite after a 520 °C anneal. The ε-carbide spectrum was quite broad and very similar to that of iron atoms with carbon near-neighbors in martensite, suggesting that the electronic states of the iron atoms coordinating the carbons in the two phases are substantially the same. It was also concluded that χ-carbide is a distinct compound, probably the Fe_5C_2 reported by Bernas et al. (1967), and not just a different physical form of Fe_3C.

The Fe–N system was also studied by Gielen and Kaplow (1967). It is in many respects similar to the Fe–C system just described. The effects of small amounts of N on the ordering of Fe_3Cr has been studied by Roy et al. (1967). They find that it results in a retention of a small amount of a new nonmagnetic phase. Marcus et al. (1966) showed that the ME can be used to determine the retained and reverted austenite in stainless steel (17–7 PH). In 18 Ni–8–Co–5–Mo maraging steels they found Fe_2Mo precipitate on annealing but no austenite.

Some years ago, it was reported that plastic deformation destroys the ME in stainless steel (Belozerskiy and Nemilov, 1963; Kortov et al., 1968). It was suggested by these authors that the reappearance of the absorption upon annealing may be attributed to the restoration of the crystal structure. It is known, however, that crystal structure is not required for the observation of the ME. The recoil-free fraction is determined entirely by the mean square amplitude of the thermal vibrations, which is not likely to be much different in a rolled foil than in an annealed one. The most likely interpretation of the 'disappearance' of the ME is that the alloy has become weakly magnetic due to a stress-induced martensite transformation, so that the absorption is spread out over the range -0.5 to $+0.5$ cm/s by inhomogeneous magnetic hfs. This can be seen in careful measurements by Kocher (1965) on various stainless steels. This point of view has also been expressed recently by Lagunov et al. (1969).

4.1.3. HYDROGEN IN PALLADIUM AND NICKEL

It is well known that palladium takes hydrogen into solid solution. The homogeneous fcc α-phase extends to about 5 at. % H (Pearson, 1958). At higher concentration, the cubic β-phase which at saturation corresponds to composition $PdH_{0.7}$ appears. The lattice constant of β-PdH is 3.7 % greater than that of α-PdH. At room temperature, the H^+ ions occupy the octahedral interstitial sites of the fcc Pd lattice. These sites constitute a second fcc lattice which together with the first makes up the NaCl structure.

In the Pd–H system there is no isotope available for a direct ME study of electronic effects on the Pd band structure. However, a number of studies have been reported in the Pd–Fe–H system (Bemski et al., 1965; Jech and Abeledo, 1967; Phillips and Kimball, 1968). The latter gives a detailed account of dilute (2 and 5 at. % Fe) alloys saturated with hydrogen. The results clearly show the two-phase nature of the alloys containing hydrogen. It is interesting to note that the β-phase orders magnetically and exhibits a ^{57}Fe hfs magnetic field practically identical to that in the α-phase. This is good evidence that the electronic structure of the iron is little affected by the hydrogen, i.e. the filling of the Pd d-band is not accompanied by similar change at the iron. The Curie temperatures of the β-phases are ≈ 0.05 of those of the corresponding α-phase. In fact, they fall well below the values computed from the Curie point of iron on the assumption of a simple dilution behavior.

The IS in the β-phase was found to be 0.0045 cm/s larger than that of the α-phase. The sign of the shift indicates decreased s-electron density at the nucleus such as could arise from the volume expansion of the β-phase. The change in IS with atomic volume ($V \Delta IS/\Delta V$) for Fe in Pd as determined by high pressure experiments is 0.06 cm/s (Ingalls et al., 1967), less than half of that in α-iron. For an 11 % increase in atomic volume, the isomer shift should then increase by 0.007 cm/s. This is in reasonable agreement with the measured value of 0.0045 cm/s and tends to confirm that the volume change is the major source of the observed shift. Decreased s-electron density could of course also result from the filling of the d-band but would be counteracted by the simultaneous filling of the s-band. The observation that the hfs effective field is the same in α- and β-phase is strong evidence, however, that the d-electron localization has not changed appreciably. The fact that the observed IS can be accounted for semiquantitatively on the basis of the lattice expansion produced by the addition of hydrogen lends support to the simple assumption that the s-electron density is inversely proportional to the volume per iron atom.

In more concentrated alloys (11 and 15 at. % Fe), Jech and Abeledo (1967) found no evidence for the formation of β-phase indicating that it is suppressed by the presence of iron. Their results show a 30% decrease in Curie temperature without any change in hfs effective field. The IS changes were attributed to lattice expansion.

The Pd–H system has also been studied using the ME of [119]Sn (Chekin and Naumov, 1966). They used an alloy containing 1.4 at.% tin, enriched to 66.3% in the isotope [119]Sn. This is well within the limit of solid solubility which has been variously placed at 10 and 26 at.%. They report the IS as a function of the hydrogen content and show that it is qualitatively similar to that of [119]Sn in the isoelectric Pd–Ag alloys. The Pd–Ag system forms a continuous fcc solid solution with a 1.7% increase in lattice constant at 40 at.% Ag. It should be noted that the total observed shift is a small fraction of the experimental linewidth (less than 10%) so that the isomer shifts of [119]Sn in the α- and β-phases could not be resolved. The reported continuous variation of the shift with hydrogen concentration may thus be an artifact of the data analysis. (It is conceivable, of course, that Sn inhibits the formation of the β-phase.) At 40 at.% H (the highest concentration examined by Chekin and Naumov), $\frac{2}{3}$ of the palladium should be in the β-phase which has a lattice constant 3.7% greater than α-Pd. Since the α- and β-phase lines are not resolved, the average line position will correspond to an effective expansion of 2.4%. Comparison of the shift in the Pd–H alloys with that in the isoelectronic Pd–Ag alloys shows qualitative agreement with that expected on the basis of the relative sizes of the lattice expansions. However, the sign of the IS predicted by the naive lattice expansion model is contrary to that observed, i.e. lattice expansion here leads to increased IS which corresponds (in [119]Sn) to increased s-electron density. The authors conclude that for Sn dissolved in Pd the change in lattice constant is not of decisive importance in determining the change in IS due to the introduction of hydrogen.

Mössbauer experiments on the pressure dependence of the IS in [119]Sn by Möller (1968) have shown, however, that the pressure shift is also contrary to that expected on the assumption that compression will increase the s-electron density at the nucleus. In metallic Sn as well as in Pd (12 wt.% Sn), he finds decreased IS with increasing pressure. This is attributed to screening effects of the 5p on the 5s electrons. In the light of these results the change of the IS in [119]Sn with introduction of hydrogen can also be attributed to the lattice expansion, although the inverse scaling of s-electron density with lattice volume is not valid in this case.

It is less well known that fcc nickel takes hydrogen into solid solution to about the same extent as palladium, provided the hydrogen is supplied in atomic form or at high pressure. The solubility of hydrogen is determined by the ability of the almost filled d-shell to take up the electron of the hydrogen atom.

The study of the effects of H in Ni could be carried out with the ME of ^{61}Ni. This interesting experiment, which should show the effects on IS and magnetic hfs interaction of filling the d-shell of nickel has not yet been carried out. This system has, however, been studied indirectly using ^{57}Co in Ni as a ME source (Wertheim and Buchanan, 1966, 1967). The results showed quite clearly that in this system the solid solubility of H in fcc Ni is small, and that a separate nonmagnetic hydride phase forms when this limit is exceeded (fig. 4.2). It was also shown that Fe retains a magnetic moment in $NiH_{0.7}$. The IS of ^{57}Fe in $NiH_{0.7}$ relative to that of in Ni is much greater than that in Cu. The sign of the shift indicates a decrease in the s-electron density which was attributed to an increase in the d-electron localization on the Fe atoms. However, it could be due, at least in part, to the volume expansion which accompanies the formation of the hydride phase. This follows from the related work on PdH discussed above.

4.2. Properties of dilute chemical impurities in metals

We now turn to those systems in which the interest is focused on the impurity-atom itself rather than on the perturbation which it produces in the host lattice. The ME is particularly well suited for the study of chemical impurities in metals, since these experiments can be carried out with dilute radioactive sources containing the parent of the Mössbauer isotope. We have seen previously that recoil effects in metals are of minor importance because of replacement collisions and that thermal and electronic effects relax rapidly. We will here consider a few illustrative examples of studies concerned specifically with the properties of such dilute impurities in metals. Of particular interest are studies of isomer shifts, magnetic hfs and recoil-free fraction.

The IS is of the greatest interest because it provides information on the s-electron charge density at the nucleus which is not obtainable from other measurements. The ^{57}Fe isomer shift in a number of metallic hosts, mainly transition metals, have been reported over the years (Walker et al., 1961; Kerler and Neuwirth, 1962; Schiffer et al., 1964; Steyert and Taylor, 1964; Housley et al., 1964; Edge et al., 1965; Bara and Hrynkiewicz, 1966). A compilation of results by Wertheim (1964) showed systematic variations in the isomer shift of ^{57}Fe in d-group transition metals. The results of a definitive

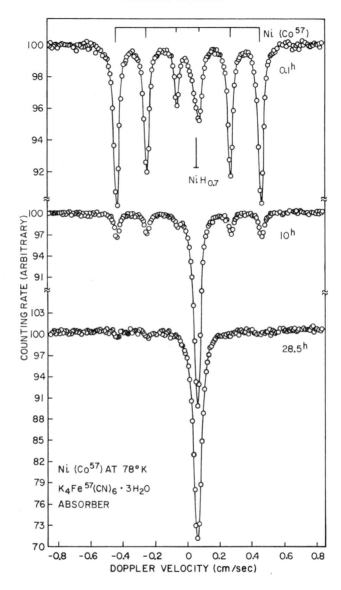

Fig. 4.2. Formation of β-phase nickel hydride with electrolytic introduction of hydrogen into a thin nickel foil. The hfs is due to ^{57}Fe resulting from the decay of ^{57}Co which was plated onto nickel and then diffused in. The six-line spectrum is that of ^{57}Fe in metallic nickel, the single line near 0.05 cm/s is due to ^{57}Fe in $NiH_{0.7}$. (From Wertheim and Buchanan, 1967.)

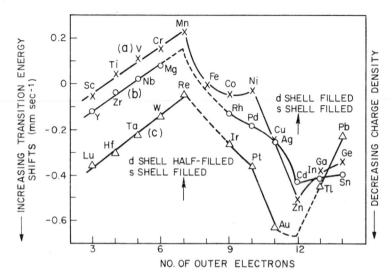

Fig. 4.3. Isomer shifts of ^{57}Fe in the 3, 4 and 5d group transition metals series plotted against the number of outer electrons (which are distributed between the nd, $(n + 1)$s and $(n + 1)$p shells) for various hosts. (From Qaim, 1967.)

study by Qaim (1967), including most of the 3, 4 and 5d-group elements, is shown in fig. 4.3. The systematic variation of the IS with filling of d- and s-shells is striking, but a detailed theoretical interpretation has not yet been given. It should be noted that there is no systematic variation with lattice parameter, although pressure experiments show a simple variation of IS with atomic volume in these systems (Edge et al., 1965; Moyzis et al., 1968).

Related results have also been obtained with ^{119}Sn by Bryukhanov et al. (1963, 1964a,b) and interpreted by Delyagin (1967) who showed that there is a correlation between the matrix compressibility and the impurity IS (see also Cordey-Hayes and Harris, 1967, and Bykov et al., 1968).

Isomer shifts of ^{197}Au in metallic hosts have been reported by Roberts and Thomson (1963), Barrett et al. (1963), Grant et al. (1964), Roberts et al. (1965) and Patterson (1967) (table 4.1). Increasing shift corresponds to increasing s-electron density. The large but uniform shift in the four 3d-group transition metals as well as in other chemically related groups is striking. It has been shown by Barrett et al. (1963) that there is a good correlation between the electronegativity of the host and the ^{197}Au isomer shift, as though the electron donating tendency of the host simply increases the s-

TABLE 4.1

Isomer shifts of [197]Au in various host lattices

Host	IS (cm/s)
Au	0.000
Pt	0.120
Ag	0.13
Se	0.18
Te	0.19
Pd	0.24
Si	0.33
Zn	0.34
Cu	0.42
Sn	0.43
Ge	0.45
Mn	0.508
Ni	0.525
Be	0.56
Co	0.563
Fe	0.577
Mg	0.63
Li	0.74
Al	0.76
Y	0.76
Ca	0.87

electron density on the gold impurity atom (fig. 4.4). Delyagin (1966) has shown that these data can also be interpreted in terms of the matrix compressibility (see also Chekin, 1968). Roberts et al. (1965) have shown a correlation between the IS and the residual electrical resistivity of gold alloys. The connection between these two properties is made through a theoretical model which uses the residual resistivity and the Friedel sum rule to specify the asymptotic wave functions at the Fermi level. A pseudopotential is then used to continue the wave function to the gold nucleus at the origin. The results of this analysis have been nicely confirmed by pressure experiments on pure gold (Roberts et al., 1969).

A linear dependence of IS on host metal matrix electronegativity has also been found for [125]Te in Pd, Ag, Au, Cu, Sn, Mg, Pb and In (Kuzmin et al., 1968). The sign of the change is opposite to that of [197]Au, indicating that p- rather than s-electrons are removed from Te by the more electronegative metals.

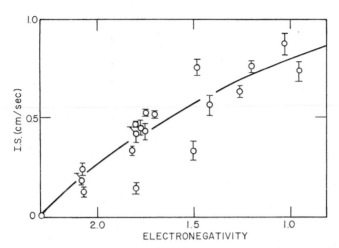

Fig. 4.4. Isomer shifts of ^{197}Au in metallic solid solution plotted against host electro-negativity. The IS reference is metallic gold itself. See table 4.1 for numerical values of isomer shifts plotted here. (From Barrett et al., 1963.)

These examples indicate that the ME isomer shift can give substantial information on the electronic properties of impurity atoms in metals. Qua-drupolar effects in metals have been systematically studied by Qaim (1969). In hcp metals he finds the QS of ^{57}Fe to be proportional to the c/a ratio of the host.

The magnetic hfs of impurity atoms in ferro- and antiferromagnetic metals has been a major area of interest because it has made possible the determination of excited state nuclear magnetic moments. From the point of view of solid state physics, the interesting properties are the atomic magnetic moment and the hfs effective magnetic field. Such fields have been measured with a variety of nuclear physics techniques; see Shirley and Westenbarger (1965), or Frankel et al. (1965) for compilations. Mössbauer effect measurements were first carried out with ^{57}Fe in the ferromagnetic metals (Wertheim, 1960) and have been carried out with rare-earth elements, e.g. ^{161}Dy in Gd (Lukashevich et al. 1966) and ^{151}Eu in Yb (Hüfner and Wernick, 1968). In the case of a transition metal impurity, H_{eff} is largely a property of the d-shell of the impurity. Of greater interest are the H_{eff} experience by nontransition elements like tin or gold where the field arises from polarized conduction electrons. Mössbauer effect measurements of the hfs magnetic fields at ^{119}Sn nuclei in ferromagnetic metals and alloys have been reported by Boyle et al. (1960), Street and Window (1966), Jain and

Cranshaw (1967), Balabanov and Delyagin (1967) and Balabanov et al. (1968), measurements of ^{197}Au in transition metals by Roberts and Thomson (1963), Grant et al. (1964) and Patterson (1967), and of ^{129}I in iron by de Waard and Drentje (1966). Such measurements have been made with a number of other ME isotopes.

Of equal interest are the properties of dilute transition metal impurities in nonmagnetic hosts, e.g. Fe in Au, Cu and Ti. These three cases provide examples of the three possible types of behavior of the magnetic moment of a transition metal impurity. Iron in Au has a magnetic moment at the lowest temperature studied; iron in Cu loses its moment due to the formation of a spin compensated state at low temperature; iron in Ti has no moment at any temperature studied.

Measurements of H_{eff} as a function of external magnetic field at low temperature in Cu have clearly shown the formation of the spin compensated state, or Kondo effect (Frankel et al., 1965; Schwartz et al., 1968; see also Kitchens et al., 1965, whose work preceded the discovery of the Kondo effect).

Dilute Fe in Au, on the other hand, has a magnetic moment and orders magnetically (Violet and Borg, 1966, 1967). More recently, Ridout (1969) has shown that the spectra of iron atoms with distinct near-neighbor configurations can be resolved in dilute gold alloys. He has obtained the dependence of the IS and QS on the number of iron neighbors. He also finds that magnetic hfs splitting appears first for iron atoms with the largest number of iron near-neighbors. Experiments in an external field confirmed the earlier finding that the magnetic order is antiferromagnetic (Violet and Borg, 1966; Gonser et al., 1965; Craig and Steyert, 1964).

Dilute Fe in Ti provides an example of the third type of behavior. Experiments with ^{57}Co in Ti sources (Taylor et al., 1964) as well as with ^{57}Fe in Ti absorbers (Rupp, 1970) show that the iron here has no magnetic moment. The latter author has also shown that no magnetic moment appears on the Fe in any of the stable or metastable phases up to the equiatomic composition in the $Ti_{1-x}Fe_x$ system.

One additional property of a dilute impurity atom which is accessible by the ME is the mean square vibrational displacement $\langle x^2 \rangle$, which is related to the recoil-free fraction f by eq. (1.5). The theory of the recoil-free fraction of impurity atoms was developed a.o. by Kagan and Iosilevskiy (1962), Maradudin and Flinn (1962), Dzyub and Lubchenko (1962), Lipkin (1963), Visscher (1963), and Housley and Hess (1966) (see also Mannheim and Simopoulos, 1968). A detailed discussion of lattice dynamics as related to the ME can be found in Maradudin (1966).

In the simplest of these theoretical analyses, the mean square displacements for the impurity is related to that of the host lattice by

$$\langle x^2 \rangle_{\text{imp}} = \langle x^2 \rangle_{\text{host}}, \qquad T \gg \theta_D,$$

$$\langle x^2 \rangle_{\text{imp}} = \left(\frac{M_{\text{host}}}{M_{\text{imp}}} \right)^{\frac{1}{2}} \langle x^2 \rangle_{\text{host}}, \qquad T \to 0, \tag{4.1}$$

provided the force constant between host and impurity atom is the same as that between host atoms.

Careful measurements have been made for [57]Fe in Be, Cu, W and Pt by Schiffer et al. (1964), in Au, Cu, Ir, Pd, Pt, Rb and Ti by Steyert and Taylor (1964), in Be and Cu by Housley et al. (1964), in Ni by Howard and Dash (1967), in Cu and Au by Housley et al. (1968), and in Cu, V and Ti by Moyzis et al. (1968). Similar measurements with [119]Sn in V have been made by Bryukhanov et al. (1963), in Au, Pt and Tl by Bryukhanov et al. (1964a) and in binary alloys by Bryukhanov et al. (1964b). At high temperature, the effects of anharmonicity were detected by Steyert and Taylor (1964) and Housley et al. (1968). The effects of pressure on f were obtained by Moyzis et al. (1968). In the case of Fe in In it has recently been shown on the basis of the large, weakly temperature dependent f that the iron is likely to be interstitial (Flinn et al., 1967), rather than substitutional as had been earlier suggested by Steyert and Craig (1962).

The work cited above provides examples of the use of the ME to study properties of dilute chemical impurity atoms in metals. While this falls into the subject of crystalline defects, much of it was not primarily motivated by the desire to contribute to that area of solid state physics. We have therefore been content to sketch rather broadly the type of information which can be obtained, provided some examples and references to other review articles.

4.3. Order–disorder transformations and precipitation

In the Fe–Ni alloy system continuous solid solution is found in the fcc γ-phase at high temperature (Pearson, 1958). (The bcc α-phase extends to ≈ 6 at. $\%$ Ni at 400 °C.) Superlattice ordering takes place in the vicinity of the composition Ni_3Fe below 506 °C. In the ordered state, three of the four simple cubic lattices which make up the fcc lattice are occupied by nickel. In the disordered state, the site occupancy is random. The transition to the ordered state by annealing has been studied in Ni_3Fe using the ^{57}Fe Möss-bauer effect by Zinn et al. (1963) and Heilmann and Zinn (1967). Although resolved neighbor effects were not obtained, they were able to obtain the short-range order parameter from an analysis of the Mössbauer line shape on the assumption that H_{eff} is determined by he first two coordination shells.

The production of short range order by irradiation of FeNi (1 : 1) in a magnetic field has been studied by Gros and Pebay-Peyroula (1964) who used electron bombardment, and by Gros et al. (1968) who used neutrons. In the former case it was shown that radiation results in the growth of a new phase with large magnetic anisotropy and a well resolved Mössbauer spectrum exhibiting quadrupole as well as magnetic hfs.

Grain boundary diffusion of ^{57}Fe in Ni foils has been observed by Bokshtein et al. (1968). After 300 h at 700 °C, they find two distinct Möss-bauer spectra corresponding to iron-rich and nickel-rich solid solutions. After 3 h at 1350 °C, only a single nickel-rich phase remains.

In the Al–Fe alloy system there is a wide range of solid solutions of Al in Fe in which the FeAl and Fe_3Al superlattices are found (Pearson, 1958). The FeAl superlattice has an ordered cubic B2 type of structure. It does not exhibit magnetic order. In a Mössbauer study, Wertheim and Wernick (1967) found that the annealed, ordered stoichiometric FeAl ex-hibits a single, slightly broadened Mössbauer line. The broadening may be due to antiphase domain boundaries or other imperfections in the long range order. Material that is crushed and not reannealed exhibits an entirely dif-ferent spectrum in which $\approx 30\%$ of the iron atoms exhibit magnetic hfs in the form of a broad background absorption (fig. 4.5). It is well established that the major effect of deformation in this system is due to slip, which introduces antiphase domain boundaries. The effect of these on the iron atom environment is to introduce two iron atoms into the near-neighbor coordination shell of Al atoms next to the slip plane. It was also verified by experiments on iron-rich FeAl that the introduction of Fe into Al sites results in regions which exhibit magnetic order at low temperature. Similar results have also been obtained by Huffman and Fisher (1967). See also Preston et al. (1966) for a study of the Fe–V system.

In Fe_3Al, the transition from a random occupancy bcc structure at high temperature to pseudo-ordered B2 structure below 800 °C and finally to the fully ordered DO_3 structure has been studied by Cser et al. (1967). For earlier Mössbauer studies of this system see Ono et al. (1962) and Johnson et al. (1963); for similar work in Fe_3Si see Stearns (1963). Cser et al. find

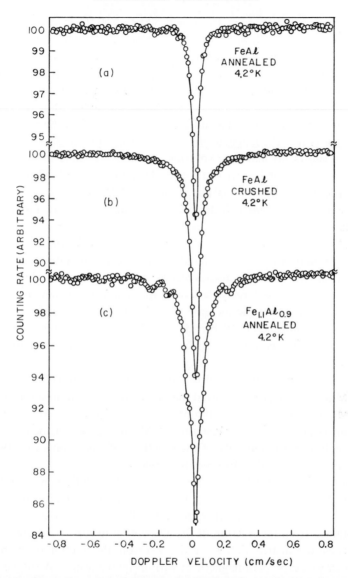

Fig. 4.5. Mössbauer absorption spectra at 4.2 K of (a) annealed FeAl, (b) crushed FeAl and (c) annealed $Fe_{1.1}Al_{0.9}$. The data in (a) are fitted in with a single Lorentzian of width 0.043 cm/s at 0.024 cm/s; those in (b) by two Lorentzians of widths 0.056 and 0.385 cm/s located at 0.024 and 0.006 cm/s, respectively. The broad line contains 30% of the total area and is due to weakly magnetic iron atoms with other iron atoms in their near-neighbor coordination shell. The data in (c) show resolved magnetic hfs due to iron atoms with excess iron neighbors. (From Wertheim and Wernick, 1967.)

evidence for a transition in the range 800 to 850 °C in a discontinuous change in the IS. The lower temperature phase is thought to be of B2 type in which one sublattice is occupied by iron atoms and the other by half iron and half aluminum atoms. Unfortunately, the iron atoms on the two distinct sublattices do not give resolved spectra. In the range from 510 to 650 °C, a fraction of the iron atoms exhibit magnetic hfs while the rest give a single paramagnetic line. The difference in IS as well as the fixed ratio of areas suggests that the magnetically ordered iron is that in D-sites while that in A-sites shows no magnetic splitting. A sudden change in both the hfs effective field and the isomer shift of the D-site iron at 600 °C is attributed to the ordering of the D-sublattice so that the DO_3 structure is attained. At the same time, the magnetic order becomes antiferromagnetic. Below 500 °C the A-sublattice also becomes magnetically ordered.

In Au_4Mn, Cohen et al. (1969), using the ME of ^{197}Au, have observed the transformation from the ordered to the disordered state by cold work. Comparison with samples disordered by quenching showed that an 80% reduction in thickness largely destroyed the long range order. In Au_4V, it was demonstrated that the increase in linewidth in the disordered material is due to inhomogeneous IS broadening. This mechanism has previously been invoked to account for the large linewidth of ^{57}Fe in nonmagnetic stainless steel.

The ME has also been used to study the precipitation of new phases in solids. Marcus et al. (1967) showed that annealing of random solid solutions of Fe–Mo at 650 °C results in the precipitation of Fe_2Mo. At 550 °C, precipitation is preceded by a clustering of Mo atoms which manifests itself in a reduction of the intensity of the spectrum of iron atoms with two Mo neighbors before the Fe_2Mo spectrum appears.

The precipitation of cobalt from Cu–Co solid solutions was studied by Nasu et al. (1967), who used a 2% Co alloy containing the isotope ^{57}Co as the source in a Mössbauer experiment. The fraction of cobalt precipitated was obtained from the fraction of the absorption in superparamagnetic hfs. The particle size distribution was estimated from the fraction of the precipitate which was superparamagnetic at various temperatures. A simple model was used to compute the critical volume corresponding to these temperatures.

The question of the mechanism by which the initial states of phase separation takes place was investigated by Nagarajan and Flinn (1967) in the Cu–Ni–Fe system. They found no evidence for the 'spinodal mechanism', i.e. the growth of small composition fluctuations of large spatial extent. Instead, the Mössbauer spectra showed the formation of precipitate of the final equilibrium composition even during the early stages of decomposition. Similar conclusions were also reached by Krogstad et al. (1963) in a study of the transitions to the ordered state in Ni_3Mn. Mössbauer spectra obtained with small amounts of ^{57}Co introduced into nearly stoichiometric Ni_3Mn show in partially ordered material the coexistence of magnetic and non-magnetic phases clearly favoring the classical nucleation and growth process.

The preceding discussion has summarized applications of the Mössbauer effect in the study of various types of defects in metals and alloys. The Mössbauer effect offers a convenient technique for the study of precipitation because it is capable of detecting minor constituents as well as very small particles. The effect of such defects as antiphase domain boundaries have been recognized in Mössbauer effect spectra. Order–disorder transitions have been repeatedly studied. Lastly, it has been shown that the effects of isolated impurity atoms on the magnetic properties of the host lattice can be resolved in magnetic systems. All of these applications are based on changes in the isomer shift or in the electric quadrupole and magnetic hfs splitting.

References

Balabanov, A. E. and N. N. Delyagin, 1967, Fiz. Tverd. Tela **9**, 1899; Soviet Phys. Solid State, English Transl. **9** (1968) 1498.

Balabanov, A. E., N. N. Delyagin, A. L. Yerzinkyan, V. P. Parfenova and V. S. Shpinel, 1968, Zh. Eksperim. i Teor. Fiz. **55**, 2136; Soviet Phys. JETP, English Transl. **28** (1969) 1131.

Bara, J. and A. Z. Hrynkiewicz, 1966, Phys. Stat. Sol. **15**, 205.

Barrett, P. H., R. W. Grant, M. Kaplan, D. A. Keller and D. A. Shirley, 1963, J. Chem. Phys. **39**, 1035.

Belozerskiy, G. N. and Yu. A. Nemilov, 1963, Fiz. Tverd. Tela **5**, 3350; Soviet Phys. Solid State, English Transl. **5** (1964) 2457.

Bemski, G., J. Danon, A. M. de Graaf and X. A. da Silva, 1965, Phys. Letters **18**, 213.

Bernas, H. and I. A. Campbell, 1966, Solid State Commun. **4**, 577.

Bernas, H., I. A. Campbell and R. Fruchart, 1967, J. Phys. Chem. Solids **28**, 17.

Bokshtein, B. S., Yu. B. Voitkovskiy, A. A. Zhukovitskiy and G. N. Pautkina, 1968, Fiz. Tverd. Tela **10**, 3699; Soviet Phys. Solid State, English Transl. **10** (1969) 2940.

Boyle, A. J. F., D. St. P. Bunbury and C. Edwards, 1960, Phys. Rev. Letters **5**, 553.

Bryukhanov, V. A., N. N. Delyagin and Yu. Kagan, 1963, Zh. Eksperim. i Teor. Fiz. **45**, 1372; Soviet Phys. JETP, English Transl. **18** (1964) 945.

Bryukhanov, V. A., N. N. Delyagin and Yu. Kagan, 1964a, Zh. Eksperim. i Teor. Fiz. **46**, 825; Soviet Phys. JETP, English Transl. **19** (1964) 564.

Bryukhanov, V. A., N. N. Delyagin and V. S. Shpinel, 1964b, Zh. Eksperim. i Teor. Fiz. **47**, 2085; Soviet Phys. JETP, English Transl. **20** (1965) 1400.

Bykov, V. N., L. N. Krizhanskiy, B. I. Rogozev, I. I. Rudnev and A. E. Fedorovskiy, 1968, Fiz. Tverd. Tela **10**, 2869; Soviet Phys. Solid State, English Transl. **10** (1969) 2267.

Campbell, I. A., 1966, Proc. Phys. Soc. London **89**, 71.

Chekin, V. V., 1968, Zh. Eksperim. i Teor. Fiz. **54**, 1829; Soviet Phys. JETP, English Transl. **27** (1968) 983.

Chekin, V. V. and V. G. Naumov, 1966, Zh. Eksperim. i Teor. Fiz. **51**, 1048; Soviet Phys. JETP, English Transl. **24** (1967) 699.

Christ, B. W. and P. M. Giles, 1968, J. Metals **20**, A51.

Cohen, R. L., J. H. Wernick, K. W. West, R. C. Sherwood and G. Y. Chin, 1969, Phys. Rev. **188**, 684.

Collins, M. F. and G. G. Low, 1964, J. Phys. Radium **25**, 596.

Collins, M. F. and G. G. Low, 1965, Proc. Phys. Soc. London **86**, 535.

Cordey-Hayes, M. and I. R. Harris, 1967, Phys. Letters **24A**, 80.

Craig, P. P. and W. A. Steyert, 1964, Phys. Rev. Letters **13**, 802.

Cranshaw, T. E., C. E. Johnson, M. S. Ridout and G. A. Murray, 1966, Phys. Letters **21**, 481.

Cser, L., J. Ostanevich and L. Pal, 1967, Phys. Stat. Sol. **20**, 581, 591.

Delyagin, N. N., 1966, Fiz. Tverd. Tela **8**, 3426; Soviet Phys. Solid State, English Transl. **8** (1967) 2748.

De Waard, H. and S. A. Drentje, 1966, Phys. Letters **20**, 38.

Dzyub, I. P. and A. F. Lubchenko, 1962, Dokl. Akad. Nauk SSSR **147**, 584; Soviet Phys. Dokl., English Transl. **7** (1963) 1027.

Edge, C. K., R. Ingalls, P. Debrunner, H. G. Drickamer and H. Frauenfelder, 1965, Phys. Rev. **138**, A729.

Flinn, P. A., U. Gonser, R. W. Grant and R. M. Housley, 1967, Phys. Rev. **157**, 538.

Frankel, R. B., J. Huntzicker, E. Matthias, S. S. Rosenblum, D. A. Shirley and N. J. Stone, 1965, Phys. Letters **15**, 163.

Genin, J. M. and P. A. Flinn, 1966, Phys. Letters **22**, 392.

Gielen, P. M. and R. Kaplow, 1967, Acta Met. **15**, 49.

Gonser, U., et al. 1965, J. Appl. Phys. **36**, 2124.

Grant, R. W., M. Kaplan, D. A. Keller and D. A. Shirley, 1964, Phys. Rev. **133**, A7062.

Gros, Y. and J. C. Pebay-Peyroula, 1964, Phys. Letters **13**, 5.

Gros, Y., J. Pauleve, D. Dautreppe and J. C. Pebay-Peyroula, 1968, Compt. Rend. **266**, 1199.

Heilmann, A. and W. Zinn, 1967, Z. Metallk. **58**, 113.

Housley, R. M. and F. Hess, 1966, Phys. Rev. **146**, 517.

Housley, R. M., J. G. Dash and R. H. Nussbaum, 1964, Phys. Rev. **136**, A464.

Housley, R. M., F. Hess and T. Sinnema, 1968, Solid State Commun. **6**, 375.

Howard, D. G. and J. G. Dash, 1967, J. Appl. Phys. **38**, 991.

Hüfner, S. and J. H. Wernick, 1968, Phys. Rev. **173**, 448.

Huffman, G. P. and R. M. Fisher, 1967, J. Appl. Phys. **38**, 735.

Ingalls, R., H. G. Drickamer and G. de Pasquali, 1967, Phys. Rev. **155**, 165.

Ino, H., T. Moriya, F. E. Fujita and Y. Maeda, 1967, J. Phys. Soc. Japan **22**, 346.

Ino, H., T. Moriya, F. E. Fujita, Y. Maeda, Y. Ono and Y. Inokuti, 1968, J. Phys. Soc. Japan **25**, 88.

Jain, A. P. and T. E. Cranshaw, 1967, Phys. Letters **A25**, 425.

Jech, A. E. and C. R. Abeledo, 1967, J. Phys. Chem. Solids **28**, 1371.

Johnson, C. E., M. S. Ridout and T. E. Cranshaw, 1963, Proc. Phys. Soc. London **81**, 1079.

Kagan, Yu. and Ya. A. Iosilevskiy, 1962, Zh. Eksperim. i Teor. Fiz. **42**, 259; Soviet Phys. JETP, English Transl. **15** (1962) 182.

Kerler, W. and W. Neuwirth, 1962, Z. Physik **167**, 176.

Kitchens, T. A., W. A. Steyert and R. D. Taylor, 1965, Phys. Rev. **138**, A467.

Kocher, C. W., 1965, Phys. Letters **14**, 287.

Kortov, V. S., R. I. Mints and Yu. N. Sekisov, 1967, Fiz. Tverd. Tela **9**, 2755; Soviet Phys. Solid State, English Transl. **9** (1968) 2169.

Krogstad, R. S., R. W. Moss and V. Vali, 1963, Phys. Letters **4**, 44.

Kuzmin, R. N., A. A. Opalenko and V. S. Shpinel, 1968, Zh. Eksperim. i Teor. Fiz. Pis'ma **8**, 455; Soviet Phys. JETP Letters, English Transl. **8** (1969) 279.

Lagunov, V. A., V. I. Polozenko and V. A. Stepanov, 1969, Fiz. Tverd. Tela **11**, 238; Soviet Phys. Solid State, English Transl. **11** (1969) 191.

Lewis, S. J. and P. A. Flinn, 1968, Phys. Stat. Sol. **26**, K51.

Lipkin, H. J., 1963, Ann. Phys. N.Y. **23**, 28.

Low, G. G. and M. F. Collins, 1963, J. Appl. Phys. **34**, 1195.

Lukashevich, I. I., V. V. Sklyarevskiy, K. P. Aleshin, B. N. Samoilov, E. P. Stepanov and N. I. Filippov, 1966, Zh. Eksperim. i Teor. Fiz. Pis'ma **3**, 81; Soviet Phys. JETP Letters, English Transl. **3** (1966) 50.

Mannheim, P. D. and A. Simopoulos, 1968, Phys. Rev. **165**, 845.

Maradudin, A. A., 1966, Solid State Phys. **18**, 273.

Maradudin, A. A. and P. A. Flinn, 1962, Phys. Rev. **126**, 2059.

Marcus, H. L. and L. H. Schwartz, 1967, Phys. Rev. **162**, 259.

Marcus, H. L., L. H. Schwartz and M. E. Fine, 1966, Am. Soc. Metals Trans. Quart. **59**, 468.

Marcus, H. L., M. E. Fine and L. H. Schwartz, 1967, J. Appl. Phys. **38**, 4750.

Möller, H. S., 1968, Z. Physik **212**, 107.

Moriya, T., H. Ino, F. E. Fujita and Y. Maeda, 1968, J. Phys. Soc. Japan **24**, 60.

Moyzis, J. A., G. de Pasquali and M. G. Drickamer, 1968, Phys. Rev. **172**, 665.

Nagarajan, A. and P. A. Flinn, 1967, Appl. Phys. Letters **11**, 120.

Nasu, S., T. Shinjo, Y. Nakamura and Y. Murakami, 1967, J. Phys. Soc. Japan **23**, 664.

Ono, K., Y. Ishikawa and A. Ito, 1962, J. Phys. Soc. Japan **17**, 1747.

Patterson, D. O., 1967, Thesis, unpublished.

Pearson, W. B., 1958, A handbook of lattice spacings and structures of metals and alloys, Vol. 4 (Pergamon Press, London).

Phillips, W. C. and C. W. Kimball, 1968, Phys. Rev. **165**, 401.

Preston, R. S., D. J. Lam, M. V. Nevitt, D. O. van Ostenburg and C. W. Kimball, 1966, Phys. Rev. **149**, 440.

Qaim, S. M., 1967, Proc. Phys. Soc. London **90**, 1065.

Qaim, S. M., 1969, J. Phys. C2, 1434.

Ridout, M. S., 1969, J. Phys. C2, 1258.

Roberts, L. D. and J. O. Thomson, 1963, Phys. Rev. **129**, 664.

Roberts, L. D., R. L. Becker, F. E. Obenshain and J. O. Thomson, 1965, Phys. Rev. **137**, A895.

Roberts, L. D., D. O. Patterson, J. O. Thomson and R. P. Levey, 1969, Phys. Rev. **179**, 656.

Ron, M., H. Shechter, A. A. Hirsch and S. Niedzwiedz, 1966, Phys. Letters **20**, 481.

Ron, M., A. Kidron, H. Shechter and S. Niedzwiedz, 1967, J. Appl. Phys. **38**, 590.

Ron, M., H. Shechter and S. Niedzwiedz, 1968, J. Appl. Phys. **39**, 265.

Roy, R. B., B. Solly and R. Wappling, 1967, Phys. Letters **A24**, 583.

Rubinstein, M., G. H. Stauss and M. B. Stearns, 1966, J. Appl. Phys. **37**, 1334.

Rupp, G., 1970, Z. Physik **230**, 265.

Schiffer, J. P., P. N. Parks and J. Heberle, 1964, Phys. Rev. **133**, A1553.

Schwartz, B. B., D. J. Kim, R. B. Frankel and N. A. Blum, 1968, J. Appl. Phys. **39**, 698.

Shinjo, T., F. Itoh, U. Takahi and Y. Nakamura, 1964, J. Phys. Soc. Japan **19**, 1252.

Shirley, D. A. and G. A. Westenbarger, 1965, Phys. Rev. **138**, 170.

Stearns, M. B., 1963, Phys. Rev. **129**, 1136.

Stearns, M. B., 1966, Phys. Rev. **147**, 439.

Stearns, M. B. and S. S. Wilson, 1964, Phys. Rev. Letters **13**, 313.

Steyert, W. A. and P. P. Craig, 1962, Phys. Letters **2**, 165.

Steyert, W. A. and R. D. Taylor, 1964, Phys. Rev. **134**, A716.

Street, R. and B. Window, 1966, Proc. Phys. Soc. London **89**, 587.

Taylor, R. D., W. A. Steyert and D. E. Nagle, 1964, Rev. Mod. Phys. **36**, 406.

Violet, C. E. and R. J. Borg, 1966, Phys. Rev. **149**, 540.

Violet, C. E. and R. J. Borg, 1967, Phys. Rev. **162**, 608.

Visscher, W. M., 1963, Phys. Rev. **129**, 28.

Walker, L. R., G. K. Wertheim and V. Jaccarino, 1961, Phys. Rev. Letters **6**, 98.

Wertheim, G. K., 1960, Phys. Rev. Letters **4**, 403.

Wertheim, G. K., 1964, Mössbauer effect, Principles and applications (Academic Press, New York).

Wertheim, G. K. and D. N. E. Buchanan, 1966, Phys. Letters **21**, 255.

Wertheim, G. K. and D. N. E. Buchanan, 1967, J. Phys. Chem. Solids **28**, 225.

Wertheim, G. K. and J. H. Wernick, 1967, Acta Met. **15**, 297.

Wertheim, G. K., V. Jaccarino, J. H. Wernick and D. N. E. Buchanan, 1964, Phys. Rev. Letters **12**, 24.

Zemcik, T., 1967, Phys. Letters **A24**, 148.

Zinn, W., M. Kalvius, E. Kankeleit, P. Kienle and W. Wiedemann, 1963, J. Phys. Chem. Solids **24**, 993.

COMMONLY USED ABBREVIATIONS, SYMBOLS AND FUNDAMENTAL CONSTANTS

A1.1. Abbreviations

EFG electric field gradient
epr electron paramagnetic resonance
hfs hyperfine structure
IS isomer shift
ME Mössbauer effect
MCA multichannel analyzer
nmr nuclear magnetic resonance
QS quadrupole splitting

A1.2. Symbols

E energy of gamma ray
E_R free-atom recoil energy
f recoil-free fraction
H magnetic field (also H_{eff})
I_g, I_e nuclear spin quantum number of ground and excited states
M mass of atom or nucleus
m_I magnetic quantum number
Q nuclear quadrupole moment
q principal component of electric field gradient
T temperature
U internal energy per unit mass
V volume

v	Doppler velocity
$\langle v^2 \rangle$	mean square thermal velocity
$\langle x^2 \rangle$	mean square thermal displacement along x axis
Z	nuclear charge
α	internal conversion coefficient
Γ_n	natural linewidth
δE_T	thermal shift
$\delta \langle r^2 \rangle$	change in mean square nuclear charge radius
θ_D	Debye temperature
$\lambda = hc/E$	wavelength of gamma radiation
μ	atomic magnetic moment
σ_0	cross section for resonant gamma ray absorption
$\sigma = \sigma_0 f$	
τ	nuclear lifetime

A1.3. Fundamental constants

c	velocity of light	$= 2.997925 \times 10^{10}$ cm/s
e	elementary charge	$= 4.80298 \times 10^{-10}$ e.s.u.
h	Planck's constant	$= 6.6256 \times 10^{-27}$ erg · s
$\hbar$	$h/2\pi$	$= 1.05450 \times 10^{-27}$ erg · s
μ_B	Bohr magneton	$= 9.2732 \times 10^{-21}$ erg/G
μ_N	nuclear magneton	$= 5.0505 \times 10^{-24}$ erg/G
k	Boltzmann's constant	$= 1.38054 \times 10^{-16}$ erg/K

NUCLEAR DECAY SCHEMES OF SELECTED MÖSSBAUER ISOTOPES

$_{19}K^{40}$

Fig. A2.1.

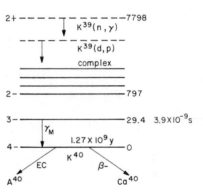

$_{26}Fe^{57}$

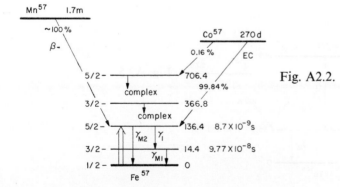

Fig. A2.2.

$_{28}$Ni^{61}

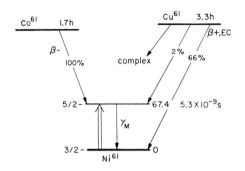

Fig. A2.3.

$_{50}$Sn^{119}

Fig. A2.4.

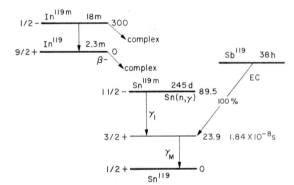

$_{52}$Te^{125}

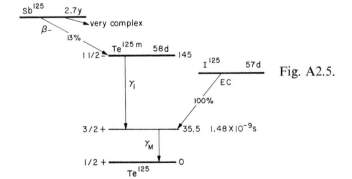

Fig. A2.5.

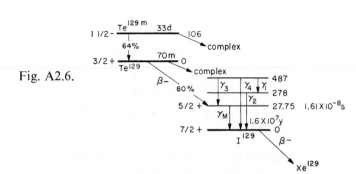

Fig. A2.6.

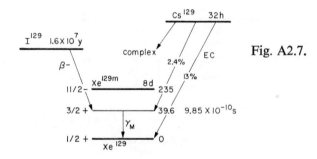

Fig. A2.7.

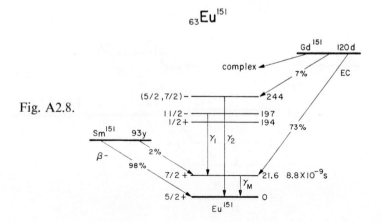

Fig. A2.8.

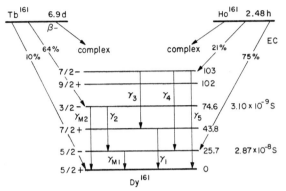

Fig. A2.9.

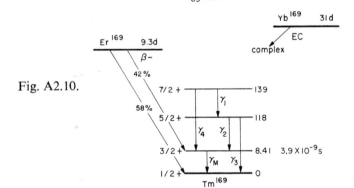

Fig. A2.10.

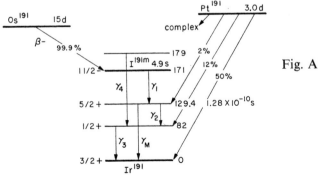

Fig. A2.11.

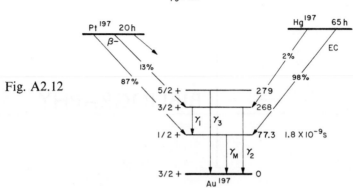

Fig. A2.12

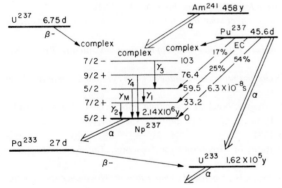

Fig. A2.13.

BIBLIOGRAPHY

A. J. F. Boyle and H. E. Hall, Mössbauer effect, *Rept. Progr. Phys.* **25** (1962) 441.

H. Frauenfelder, *The Mössbauer effect* (Benjamin, New York, 1962).

V. I. Goldanskiy, *The Mössbauer effect and its applications in chemistry* (Consultants Bureau, New York, 1964.)

G. K. Wertheim, *Mössbauer effect, Principles and applications* (Academic Press, New York, 1964).

H. Wegener, *Der Mössbauer Effekt und seine Anwendungen in Physik und Chemie* (Bibliogr. Inst. A.G., Mannheim, 1965).

K. G. Malmfors and R. L. Mössbauer, Nuclear resonance fluorescence of gamma radiation, in: K. Siegbahn, ed., *Alpha-, beta- and gamma-ray spectroscopy, Vol. II* (North-Holland, Amsterdam, 1965) 1281.

R. H. Herber, Chemical applications of Mössbauer spectroscopy, *Ann. Rev. Phys. Chem.* **17** (1966) 261.

V. I. Goldanskiy and R. H. Herber, eds., *Chemical applications of Mössbauer spectroscopy* (Academic Press, New York, 1968).

E. Matthias and D. A. Shirley, eds., *Hyperfine structure and nuclear radiations* (North-Holland, Amsterdam, 1968).

A bibliography of Mössbauer effect research is available in:
A. H. Muir, K. J. Ando and H. M. Coogan, *Mössbauer effect data index, 1958–1965* (Interscience, New York, 1966).

AUTHOR INDEX

SUBJECT INDEX

STUDY OF STRUCTURAL DEFECTS BY SPIN RESONANCE METHODS

A. HAUSMANN

and

W. SANDER

2. Physikalisches Institut der
Rheinisch-Westfälischen Technischen Hochschule
51-Aachen
Germany

PREFACE

Spin resonance methods must now be considered standard techniques in solid state physics, as are low temperature or optical absorption techniques. The field is covered by quite a number of books, both on introductory and on advanced levels. Most of the experimental equipment is commercially available. Numerous results have been published, and are summarized in many review articles which concentrate on different classes of substances. Therefore the authors feel it hard to bring forward some justification to put this new book aside the other ones. They think — as the editors do — that a new series of monographs on defects in solids would be incomplete without dealing with resonance methods. In fact, the discussion of these methods under the point of view of defect study has been a lead to the authors in the conception of this book. They hope, that it might be of some help to solid state physicists, who want to be informed about the additional and sometimes unique possibilities, which spin resonance methods contribute to the investigation of defects in solids.

Acknowledgements

The authors would like to express their appreciation to the many people who have assisted in completing this book. Special thanks go to the editors whose support and diligence made the book possible and who carefully read the manuscript. Our thanks are also due to the secretarial staff of 2. Physikalisches Institut, in particular to Mrs. Inge Muschelknautz. We thank the authors and publishers who gave their permission for the reproduction of a number of figures, and finally, the publishers of this book for their assistance in the various editorial aspects involved.

A. HAUSMANN
W. SANDER

INTRODUCTION

Aim and plan of this book

Since their invention some tenths of years ago, spin resonance methods have provided a large number of results not only in physics but also in chemistry and in biology. For the study of defects in solids, the most important of these methods is the electron spin resonance (ESR). For several defects, very detailed information has been achieved by the higher resolution electron nuclear double resonance (ENDOR) technique. Both methods are sometimes included in the term 'paramagnetic resonance'. The nuclear magnetic resonance (NMR) can be used to study defects by their distortion fields resulting in quadrupole interactions.

In part 1 of this book the basic principles of ESR and ENDOR are dealt with. Some typical examples are presented in more detail in order to show what information on defects can be drawn from the analysis of their resonance spectra. The NMR is treated rather shortly, as its application to defects is rather limited as compared to the electron resonance methods.

In part 2, results for several classes of solids are summarized. Also in this part more emphasis has been laid upon giving some representative results than on attempting an encyclopedic coverage of the matter. Therefore the representation is limited to the work done on single crystals.

Of course, this method of selecting some results for closer treatment somehow reflects the personal interests of the authors. Therefore they want to apologize to many colleagues who might think other examples more important or more instructive than those preferred here.

Part A

Basic Principles of
Spin Resonance Spectroscopy

1 | ELECTRON SPIN RESONANCE

1.1. Basic concepts

Electron spin resonance (ESR) is concerned with the investigation of resonant absorption of electromagnetic radiation by unpaired electrons or systems of unpaired electrons with total nonzero spin in a magnetic field. A system with partly filled electron orbitals will generally have a degenerate ground state, and associated with this degenerate ground state there is usually a nonzero magnetic moment. That means, such a system possesses an intrinsic magnetic moment and acts like a little magnet. In an externally applied magnetic field there are several possible orientations of the magnetic moment and associated with them distinct energy levels. Transitions between these levels give rise to resonance lines of finite width.

The ordinary resonance transitions occur between adjacent energy levels which have been separated by an externally applied magnetic field H. We first consider the simple case of unpaired electrons associated with a free radical specimen or of paramagnetic centres in which there is very little coupling of the electrons to the orbital motion. In this case, the unpaired electrons will effectively only have a spin angular momentum and therefore possess a magnetic moment equal to one Bohr magneton. The externally applied magnetic field interacts with the magnetic moments of these unpaired electrons, and the electrons will be separated into two groups. One group has a spin component lined up along the direction of the magnetic field, corresponding to a quantum number $M = +\frac{1}{2}$. The other group is lined up with the spin component antiparallel to the field and has a designation of $M = -\frac{1}{2}$. These groups also have different energies.

The energy of a magnetic moment μ aligned in a magnetic field H is equal to $-\mu H$. The magnetic moment of the unpaired electrons is given by $\frac{1}{2}g\mu_B$. Here g is the spectroscopic splitting factor, which for a free electron with no orbital momentum is equal to $g_s = 2.00229$ and for a free atom has the value given by the Landé splitting factor. The g value measures the contribution of the orbital and spin angular momentum to the magnetic moment. The Bohr magneton μ_B converts the units of angular momentum to units of magnetic moment. The energies of the two groups of electrons thus are equal to $-\frac{1}{2}g\mu_B H$ and $+\frac{1}{2}g\mu_B H$. The effect of the application of the magnetic field is to produce an energy splitting equal to $g\mu_B H$ (Zeeman effect). The basic principle of electron spin resonance is then to provide electromagnetic radiation such that

$$h\nu = g\mu_B H. \tag{1.1}$$

Then electrons can be excited from the ground level to the higher level and this can be detected by the absorption of the incoming microwave frequency (see fig. 1.1). The ESR transitions fulfill the selection rules $\Delta M = \pm 1$, whereas the nuclear spin quantum number remains unchanged.

The electromagnetic radiation not only causes electrons in the lower level to be excited to the higher level and to absorb a quantum of radiation, but it also produces stimulated emission of those in the higher level which return to the ground state, emitting a quantum of the resonance frequency. The coefficients for both processes are equal. The reason why a net absorption is detected is the fact that there is a larger number of electrons in the ground state than in the higher state. The probability for an electron in the ground state to be excited is therefore greater than for one in the higher state. From this it follows that the intensity of the absorption will depend

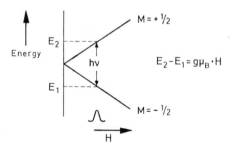

Fig. 1.1. The Zeeman splitting of an energy level

on the difference in population between the two levels. This difference is governed by the Maxwell–Boltzmann distribution law

$$\frac{N_1}{N_2} = \exp \frac{g\mu_B H}{kT}. \tag{1.2}$$

An increasing difference in population will be produced, the larger the value of the magnetic field and the lower the temperature T.

In the resonance condition (eq. 1.1), there are two linearly related variables on each side. Principally, electron spin resonance can be performed at any field strength and appropriate frequency. There is, however, a good reason, viz. the sensitivity of the spectrometer, why one normally works at as high values of the magnetic field as possible. One generally applies frequencies in the 10^{10} Hz range, the fields then being of the order of 10^3 to 10^4 G. These field strengths are still easily available in the laboratory, and for the microwave device the experience obtained with the marine radar band at 9×10^9 Hz and the airport control radar band at about 3.5×10^{10} Hz can be utilized. (For details see also section 1.11.)

The absorption of energy which gives the ESR spectrum occurs because there is a change in the magnetic moment of the ion or radical in question. What in fact we are detecting is a change in the magnetic susceptibility arising from the change in the magnetic moment. In the presence of the radio frequency field the susceptibility becomes complex,

$$\chi = \chi' - i\chi''. \tag{1.3}$$

It contains two components: an imaginary part χ'', a change in which gives rise to pure absorption, and a real part χ' which gives a pure dispersion. The absorption curve may always be derived from the dispersion curve and vice versa by use of the Kronig–Kramers equation. However, the absorption trace is mostly studied in preference to the dispersion trace, except where power saturation will tend to broaden the absorption lines. Most commercially available spectrometers can be used to detect both absorption and dispersion.

The basic principles of electron spin resonance (ESR) have been discussed in a number of books, not counting the numerous original papers. In this book, only the most important concepts are briefly outlined as far as they are of use to an experimentalist in interpreting the observed spectra. We do not consider the more fundamental problem of what are the reasons for the various interactions involved, but describe the experimental results in terms of the parameters in a 'spin Hamiltonian'.

In this chapter, we derive this spin Hamiltonian from the picture of the free ion Hamiltonian. Next we consider the modifications arising if the ions are imbedded in a solid. Discussing a simple example, we then work out what sort of information may be obtained from the spectra.

1.2. The free ion Hamiltonian

In this book, we mainly deal with the properties of isolated paramagnetic centres in an otherwise nonmagnetic crystal. The energy levels of an atom or ion incorporated in a crystalline solid of given symmetry are subject to many interactions. If we want to understand the resulting modifications of the energy levels, it is convenient to consider first the interactions within a free ion. Here the dominant contribution to the magnetism comes from the electrons. Nuclear magnetism is an order of magnitude smaller, though very important in causing the hyperfine structure (HFS) in ESR.

First we consider the complete Hamiltonian of a free atom, which represents the total energy of the system and which may be considered as the sum of all energy contributions. Of course, some contributions are more important than others, and we shall consider later the relative magnitudes of these energies. A particular part of the overall Hamiltonian is the spin Hamiltonian which was introduced by Abragam and Pryce (1951). By far it is the most useful mathematical tool in the techniques of spin resonance.

The Hamiltonian of a free atom may be considered to be the sum of the following components:

(i) H_c is the most important interaction in an atom, describing the interaction of the electrons with the nuclear charge as well as the mutual repulsion of the electrons.

(ii) H_{LS}, the next important term, is a magnetic interaction between the electron spins and the orbital momentum, the spin–orbit interaction. For Russel–Saunders coupling, it can be written as λLS, i.e. an interaction of orbital angular momentum L with electron spin S. The constant λ for a given ion can be taken from optical spectra and may be positive or negative.

(iii) H_{SS}, the mutual magnetic interaction between the dipoles (spin–spin interaction), is much weaker and is often small enough to be ignored, but it may be important in some transition elements.

(iv) If the nucleus has a spin $I \neq 0$ and an electric quadrupole moment, we have to consider terms H_N and H_Q, which describe the interaction of nuclear spins. H_N gives the dipole–dipole interaction between the nuclear moment and the magnetic moment of the electrons and the interaction of the s electrons with the nuclear spin I. H_Q describes the interaction of the electric quadrupole moment Q of the nucleus with the gradient of the electric field caused by anisotropy of the electronic charge distribution at the nucleus.

(v) In the presence of an external magnetic field, the interaction of the electrons with the field is given by H_H, the direct interaction of the nucleus with the field is H_{NH}. The term H_H is the most important one in ESR and is called the Zeeman term. The other term is extremely small and may be considered negligible in ESR. In nuclear magnetic resonance this becomes the Zeeman term and is extremely important.

The Hamiltonian of a free ion is the sum of the different terms

$$H = H_c + H_{LS} + H_{SS} + H_H + H_N + H_Q + H_{NH}. \qquad (1.4)$$

In spectroscopy, the energy is commonly given in units of cm^{-1}. The definition of this unit in spin resonance spectroscopy is

$$\frac{1}{\lambda}\,[\text{cm}^{-1}] = \frac{g\mu_B H}{hc}. \qquad (1.5)$$

For the X band, with $H = 3\,000$ G and $g = 2$, one gets from eq. (1.5) $H_H \approx 1$ cm^{-1} and $H_{NH} \approx 10^{-4}$ cm^{-1}. For the other terms in the Hamiltonian, the order of magnitude of the energies may be estimated from atomic spectra, yielding $H_c \approx 10^4$ cm^{-1}, $H_{LS} \approx 10^2$ cm^{-1}, $H_{SS} \approx 1$ cm^{-1}, $H_N \approx 10^{-2}$ cm^{-1} and $H_Q \approx 10^{-3}$ cm^{-1}.

The magnetic moment of the system is an operator which conveniently is derived by differentiating the Hamiltonian with respect to the magnetic field. In zero magnetic field it is $-\mu_B(L + 2S)$, and taking into account radiation corrections gives $-\mu_B(L + g_s S)$, where $g_s = 2.00229$. The magnetic properties of the system depend very much on the expectation value of μ in the ground state. If the state is nondegenerate, $\langle \mu \rangle$ is invariably zero in the absence of an external field. In degenerate states, we have to deal with the matrix elements of μ within the ground manifold, which generally differ from zero.

1.3. The Hamiltonian of ions in a crystalline solid

If the ions are not free but incorporated in a solid, there are strong modifications of the energy levels. The group of lowest energy levels, i.e. the ones between which microwave transitions are induced, depend in a complicated way on the particular ion, the symmetry and strength of the crystalline field, the spin–orbit coupling and the hyperfine interaction between electrons and nuclei. However, the experimental resonance data can be described by the use of a spin Hamiltonian in a fairly simple way that requires no detailed knowledge of the above effects. We shall now try to approach the concept of the spin Hamiltonian.

The interactions between the electrons and their environment are of two kinds:

(1) interactions between the like magnetic dipoles,

(2) interactions between the dipoles and the diamagnetic neighbours and their nuclei.

In this book we are only interested in isolated paramagnetic centres being present with low concentration in an otherwise nonmagnetic crystal. Therefore interaction (1) is negligible for our purpose. However, if one is studying paramagnetic salts, the first interaction may be important but can be reduced to a very small amount by using crystals diluted with a diamagnetic isomorphous salt.

The interaction of the spin and its environment is of great interest to us. The surroundings may interact with the spin in several ways:

(i) The neighbours consist of charged ions setting up strong electric fields, the symmetry of which depends on the local arrangement of the diamagnetic neighbours. The effect of this electric crystal field is to split the ground states of the magnetic ion into a number of components, the number of which and the amount of splitting depending on the symmetry of the fields and on their strengths.

(ii) If we are not dealing with single electrons but with a system of electrons having total spin $S > \frac{1}{2}$, the energy is modified by the magnetic dipole–dipole interaction of the spins, which normally is referred to as fine structure (FS).

(iii) The spin–orbit coupling in a solid further modifies the energy of the free ion and causes the g value to deviate from the free spin value g_s and in general to depend on the orientation of the external field relative to the symmetry axis of the crystal field.

(iv) Further the magnetic interactions of the electrons with magnetic moments of nearby nuclei have to be taken into account. There are also electrostatic and covalent interactions with the crystalline environment, e.g. quadrupole interaction.

(v) Another important influence of the lattice is the spin–lattice relaxation, characterized by the longitudinal relaxation time T_1. This relaxation provides the mechanism by which energy is transferred from the spin system into the lattice and thus leads to thermal equilibrium and restores it after this has been disturbed by absorption of radiation. The coupling of the spin system to the lattice can be understood in such a way that the foregoing interactions are modulated by lattice vibrations or by diffusion in the lattice depending on temperature. Together with the interactions between the various magnetic dipoles these effects are also controlling the spin–spin relaxation mechanism, characterized by the transverse relaxation time T_2 (see section 1.9).

It has already been mentioned that the crystalline environment exerts strong electric fields on the electrons thus changing their behaviour. Therefore the problem to be solved is finding the energy levels in a Stark field of certain strength and symmetry.

Assuming the simplifying concept of the static crystal field (point charge model), one calculates the effect of the electric field on the levels of the free ion by means of perturbation theory. In this picture, emphasis is put on a purely ionic description, which for most purposes is quite satisfactory for a first orientation. Often, however, one will have to take account of the partly nonlocalized nature of paramagnetic centres.

In the crystal field approach, one assumes that in the region in which the electrons move the charge density arises only from the paramagnetic ion itself, is spherically symmetric, and that the charge density associated with the rest of the crystal is zero in this region. The potential arising from the rest of the crystal without the charge satisfies Laplace's equation. It can be expanded in terms of harmonics, namely polynomials in the coordinates x, y, z. We need not take care of indefinitely high order of harmonics since the matrix elements which are calculated in perturbation theory for the electrons being in s, p, d or f orbitals vanish above some order for the various orbitals. One applies perturbation theory to calculate how the energy levels of the free ion are modified by interaction with the crystal field. This can be done by assuming that these interactions change the energy by amounts which are very much smaller than the original energy. This method can be extended to include a series of decreasing energy terms, provided that each new interaction is small compared with the previous one. When two energies have roughly equal energies, they must be taken together.

The relative order of magnitude of the crystal field term with respect to the other terms in the Hamiltonian decides at which point in the calculation the crystalline field will be used as a perturbation. The magnitude of the potential in a special case can be obtained from the optical spectra of the free ion.

All crystals have the common feature of being symmetric. This means that the potential is invariant under the operations of the symmetry group the crystal belongs to, which makes it possible to obtain the Stark splitting caused by a crystalline field of given symmetry by means of group theoretical considerations. Data for many crystal structures thus derived are tabulated in the review of Low (1960). Useful methods for carrying out the perturbation calculations have been developed by Pryce (1950) and Abragam and Pryce (1951).

Generally, the theoretical description of a paramagnetic entity is so complex that a complete solution for the energy levels and wave functions is not very practical, as one would find out that the energy levels are a complicated mixture of orbital and spin wave functions of the free ion. Commonly, ESR phenomena are limited to a set of low energy levels and these levels can be regarded as isolated, forgetting about higher levels. The magnetic behaviour of the levels in question is then practically described by a 'spin Hamiltonian'. If transitions between $2S + 1$ levels are observed experimentally, we define S as the fictitious spin of the system, corresponding to $2S + 1$ possible orientations for the magnetic dipole.

Abragam and Pryce (1951) have shown that the behaviour of a spin system can be described by a spin Hamiltonian which is a polynomial in S, and that the splitting, which may be calculated by first and higher order perturbation theory, is precisely the same as if one ignored the orbital angular momentum and replaced its effect by an anisotropic coupling between the electron spin and the external field.

The spin Hamiltonian is a shorthand description of the experimental results, containing parameters which refer to the various interactions appearing as terms of energy. These parameters have to be determined by experiment, i.e., the g factor, the crystal field splittings, the HFS parameters, the quadrupole moment and the nuclear g factor.

The formal development of the spin Hamiltonian from the actual free ion Hamiltonian is as follows: The various terms in eq. (1.4) are transformed into the appropriate angular momentum operators L, S and J. Written in a somewhat abbreviated form, this results in the spin Hamiltonian

$$H_{\text{spin}} = \sum_{i,j} [\mu_B g_{ij} H_i S_j + D_{ij} S_i S_j + A_{ij} S_i I_j + P_{ij} I_i I_j$$

$$- (g_I \mu_N + R_{ij}) H_i I_j],\tag{1.6}$$

where small terms of higher order have been neglected. The quantities g_{ij}, D_{ij}, A_{ij}, P_{ij} and R_{ij} are three dimensional symmetric second rank tensors which can be reduced to diagonal form. The indices i and j refer here to the three Cartesian coordinates x, y, z.

The interpretation of the factors in eq. (1.6) is as follows: g_{ij} is the spectroscopic splitting factor. It contains, besides g_s, the contribution of admixed higher orbital states. D_{ij} is a measure of the splitting of the ground state in a crystal field of lower than cubic symmetry. This splitting is partly caused by spin–orbit coupling and by the spin–spin contribution in an asymmetric crystal field. The splitting gives rise to the socalled fine structure (FS), and in the absence of a magnetic field is called the zero-field splitting. The term with A_{ij} expresses the magnetic hyperfine structure to which un-paired s electrons and a small spin–orbit term contribute. The term with P_{ij} gives the quadrupole interaction, whereas R_{ij} expresses the interaction of the external field with the nuclear moment.

The spin Hamiltonian contains variables which must conform to the total symmetry about the ion in the crystal. For simple symmetries, such as cubic, axial and orthorhombic symmetries, the spin Hamiltonian takes the forms

$$H_{\text{cubic}} = g\mu_{\text{B}} \boldsymbol{H} \cdot \boldsymbol{S} + A \boldsymbol{S} \cdot \boldsymbol{I} - R \boldsymbol{I} \cdot \boldsymbol{H} - g_I \mu_N \boldsymbol{H} \cdot \boldsymbol{I}, \qquad (1.7)$$

$$\begin{aligned}
H_{\text{axial}} = {}& \mu_{\text{B}} \left[g_{\|} H_z S_z + g_{\perp} (H_x S_x + H_y S_y) \right] \\
& + D \left[S_z^2 - \tfrac{1}{3} S (S + 1) \right] \\
& + A_{\|} S_z I_z + A_{\perp} (S_x I_x + S_y I_y) \\
& + P \left[I_z^2 - \tfrac{1}{3} I (I + 1) \right] \\
& - R_{\|} I_z H_z - R_{\perp} (I_x H_x + I_y H_y) \\
& - g_I \mu_N \boldsymbol{I} \cdot \boldsymbol{H}, \qquad\qquad\qquad\qquad\qquad (1.8)
\end{aligned}$$

$$\begin{aligned}
H_{\text{orthorhombic}} = {}& \mu_{\text{B}} (g_z H_z S_z + g_y H_y S_y + g_x H_x S_x) \\
& + D \left[S_z^2 - \tfrac{1}{3} S (S + 1) \right] + E (S_x^2 - S_y^2) \\
& + A_z S_z I_z + A_y S_y I_y + A_x S_x I_x \\
& + P' \left[I_z^2 - \tfrac{1}{3} I (I + 1) \right] + P'' (I_x^2 + I_y^2) \\
& - R_z H_z I_z - R_y H_y I_y - R_x H_x I_x \\
& - g_I \mu_N (H_z I_z + H_y I_y + H_x I_x). \qquad\qquad (1.9)
\end{aligned}$$

Let us discuss the first two terms in eq (1.6) a little bit more, forgetting about hyperfine and quadrupole interactions for which hold similar argu-ments. Let us assume for example $S = \tfrac{3}{2}$.

In the cubic crystal field, the magnetic properties must show cubic symmetry. The tensors are therefore isotropic and the choice of the axes is arbitrary. It is convenient to take the z direction parallel to the magnetic field, in which case the Hamiltonian reduces to the simple form

$$H = g\mu_{\mathrm{B}}HS_z. \tag{1.10}$$

The energies for the four possible levels are given by $g\mu_{\mathrm{B}}HM$, with $M = \frac{3}{2}, \frac{1}{2}, -\frac{1}{2}, -\frac{3}{2}$. There is no zero-field splitting, $D = 0$ and only one resonance transition can be observed.

When the symmetry of the magnetic ion site is tetragonal or trigonal, the g factor is no longer isotropic but must also show axial symmetry: $g_x = g_y \neq g_z$. The crystal field now separates the two doublets $M = \pm\frac{1}{2}$ and $M = \pm\frac{3}{2}$, and it is necessary to introduce a new term into the spin Hamiltonian to take account of this. It has been done by the second term in eq. (1.8).

The operator S_z^2 is diagonal, so in zero magnetic field it is easy to see that the eigenvalues are the energies

$$E_{\pm 1/2} = -D, \qquad E_{\pm 3/2} = +D, \tag{1.11}$$

i.e. the zero-field splitting is $2D$. The constant term $\frac{1}{3}S(S+1)$ is included to make the matrix of the D term traceless. It has, however, no effect on the relative positions of the levels as it shifts both doublets equally. If the magnetic field is applied parallel to the z axis, the levels occur at

$$E_{\pm 1/2} = \pm\tfrac{1}{2}g_z\mu_{\mathrm{B}}H - D, \qquad E_{\pm 3/2} = \tfrac{3}{2}g_z\mu_{\mathrm{B}}H + D. \tag{1.12a}$$

If the magnetic field is normal to the z axis, the energies are given by

$$E_{\pm 1/2} = \pm\tfrac{1}{2}g_x\mu_{\mathrm{B}}H + \tfrac{1}{2}D, \qquad E_{\pm 3/2} = \pm\tfrac{3}{2}g_x\mu_{\mathrm{B}}H - \tfrac{1}{2}D. \tag{1.12b}$$

In the orthorhombic crystal field, the g tensor has three unequal principal values $g_x \neq g_y \neq g_z$, and a second crystal field term $E(S_x^2 + S_y^2)$ is added. In this case, the energies are no longer linear functions of the magnetic field except when $g\mu_{\mathrm{B}}H \gg D, E$ and are more difficult to calculate.

To find the energies with H applied off axis is rather complex. This can be done numerically with the aid of an electronic computer.

1.4. A simplified example

As a specially simple case we now deal with ESR of an S state ion. Paramagnetic impurity centres often found in zinc oxide (ZnO) single crystals are due to divalent manganese ions substituting for divalent zinc ions.

Manganese belongs to the iron group elements having $3d^5$ configuration.

The Mn^{2+} ion has a half filled shell. According to Hund's rule, electron shells which are half filled, such as d^5 or f^7, have a ground state in which the spins of all the electrons are aligned and which is an orbital singlet or S state. Therefore the ground state of the Mn^{2+} ion is $^6S_{5/2}$, being sixfold degenerate in absence of the external field. Neither crystal field nor spin–orbit coupling alone are able to remove the sixfold degeneracy. Higher order perturbations are necessary and the expected ground state splitting will be small. In medium or weak crystal fields, the ground state remains an S state. Manganese has only one isotope, Mn^{55}, the nucleus having $I = \frac{5}{2}$ and an electric quadrupole moment.

Zinc oxide crystallizes in the wurtzite structure, C_{6v}^4, having two molecules in the primitive unit cell. The crystal field is axial, to be noticed by the polar axis c. Distortions of cubic and lower symmetry have also to be considered; however, their order of magnitude is some ten times smaller than the axial part. The crystal field in a wurtzite structure may be considered to arise from a trigonal distortion along the symmetry axis c of a cubic field that is produced by a regular tetrahedron of anions (O^{2-}) around the impurity (Mn^{2+}).

Then the spin Hamiltonian for Mn^{2+} in ZnO can be written as (see eq. 1.8)

$$
\begin{aligned}
H = {} & \mu_B \left[g_{\parallel} H_z S_z + g_{\perp} (H_x S_x + H_y S_y) \right] \\
& + D \left[S_z^2 - \tfrac{1}{3} S(S+1) \right] \\
& + \tfrac{1}{180} F \left[35 S_z^4 - 30 S(S+1) S_z^2 + 25 S_z^2 \right] \\
& + \tfrac{1}{6} a \left[S_{x'}^4 + S_{y'}^4 - \tfrac{1}{5} S(S+1)(3S^2 + 3S - 1) \right] \\
& + A_{\parallel} S_z I_z + A_{\perp} (S_y I_y + S_x I_x) \\
& - P \left[I_z^2 - \tfrac{1}{3} I(I+1) \right] \\
& - g_I \mu_N \, H \cdot I.
\end{aligned}
\tag{1.13}
$$

This Hamiltonian has been augmented with terms in F and a. Here a is the coefficient of the cubic part of the crystalline field potential and F is the fourth order term of the trigonal field. For $S > 2$ in a tetrahedral symmetry, a fourth order term in S is the lowest capable of yielding a splitting of the energy levels in zero magnetic field. The nuclear g factor is taken isotropic.

We now consider how the energy levels of the ground state may be found from a given spin Hamiltonian. If the static magnetic field is sufficiently strong, the Zeeman energy is the most important part of the Hamiltonian. Thus the direction of the magnetic field is the axis of quantization and the other terms can be regarded as a perturbation. For a discussion of the Hamiltonian we have to solve the equation

$$H |\psi_{M,m}\rangle = E_{M,m}|\psi_{M,m}\rangle. \tag{1.14}$$

Here M and m are the electron and nuclear spin quantum numbers, respectively. Because $S = \frac{5}{2}$ and $I = \frac{5}{2}$, their values are $-\frac{5}{2}, -\frac{3}{2}, \ldots, +\frac{5}{2}$. ψ represents the wave function of the spin states. Eq. (1.14) means that the energy levels are obtained by finding the eigenvalues of the Hamiltonian. The eigenfunctions give $2S + 1$ states $|S\rangle, |S-1\rangle, \cdots, |-S\rangle$. If there are zero-field splittings between the levels, then with a weak magnetic field parallel to the z axis each state is, in general, of the form

$$\psi = \alpha|S\rangle + \beta|S-1\rangle \cdots, \tag{1.15}$$

where $|\alpha|^2 + |\beta|^2 + \cdots = 1$.

One way of formulating the problem of finding the eigenvalues for a state is that values of α, β, ... must be found such that the operation $H|\psi\rangle$ reproduces the same wave function $|\psi\rangle$ multiplied by a number, this number being the required eigenvalue. The spin operators S_x, S_y and S_z must be used acting on these spin states, giving the energy matrix. The eigenvalues satisfy the relation $H\psi = E\psi$. Therefore $(H - E)\psi = 0$ and the eigenvalues are the roots of the secular determinant formed by adding $-E$ to each diagonal element.

In the case where $I \neq 0$ and where there is hyperfine interaction, each spin state $|M\rangle$ consists of $2I + 1$ states which may be rewritten as $|M,m\rangle$. The rules for operating with I_x, I_y, I_z are completely analogous to those for S_x, S_y, S_z. To begin the perturbation, one has to relate the crystal field and HFS terms to the coordinate system of the unperturbed operator (electron Zeeman term), that means to a system whose axis is parallel to the external field H. This is done by the rules of vector and tensor transformation. The higher order perturbations admix some of the higher excited states into the ground state. The value of g therefore differs from the free electron value that is actually expected for an S state. The g shift Δg is approximately $-2 (\lambda/\Delta E)^2$, where λ is the spin–orbit coupling constant and ΔE is the energy separation of the ^4P state from the $^6S_{5/2}$ state.

The values for λ and ΔE may be taken from the optical spectra. The Zeeman term will determine the position of the spectrum which will vary between two extremes characterised by $g_{\parallel}$ and $g_{\perp}$ in directions parallel and normal to the c axis, respectively. At arbitrary direction of H with an angle ϑ to the c axis, the g value is given by

$$g^2 = g_{\|}{}^2 \cos^2 \vartheta + g_{\perp}{}^2 \sin^2 \vartheta. \tag{1.16}$$

The HFS is anisotropic and is characterized by the constants $A_{\|}$ and $A_{\perp}$ for the two extremes. For greater simplicity, in the following discussion g and A are asumed to be isotropic, which in experiment is fairly well satisfied. Diagonalisation of the Hamiltonian yields

$$
\begin{aligned}
\boldsymbol{H} = {}& g\mu_{\mathrm{B}} H S_z + \tfrac{1}{2} D \left[S_z{}^2 - \tfrac{1}{3} S (S + 1) \right] \left[3 \cos^2 \vartheta - 1 \right] \\
& + D \left(S_z S_x + S_x S_z \right) \cos \vartheta \sin \vartheta + \tfrac{1}{4} D \left(S_+{}^2 + S_-{}^2 \right) \sin^2 \vartheta \\
& + A \, \boldsymbol{I} \cdot \boldsymbol{S} \\
& + \tfrac{7}{36} F \left\{ S_{z'}{}^4 - \tfrac{1}{7} \left[6 S (S + 1) - 5 \right] S_{z'}{}^2 \right. \\
& \left. + \left[\tfrac{3}{35} S (S + 1) \right] \left[S (S + 1) - 2 \right] \right\} \\
& + \tfrac{1}{6} a \left[S_{x'}{}^4 + S_{y'}{}^4 + S_{z'}{}^4 - \tfrac{1}{5} S (S + 1) (3 S^2 + 3 S - 1) \right] \\
& + P \left[I_z{}^2 - \tfrac{1}{3} I (I + 1) \right] \\
& - g_I \mu_N \, \boldsymbol{H} \cdot \boldsymbol{I}. \tag{1.17}
\end{aligned}
$$

For simplification, the terms with a, F and P have been written in the old coordinate system as they are not so important for the further discussion.

To get a survey of the spectrum, we first consider a Hamiltonian consisting of the first two and of the fifth term of eq. (1.17). First-order perturbation treatment gives the energy levels

$$E_{M,m} = g\mu_{\mathrm{B}} H M + \tfrac{1}{2} D (3 \cos^2 \vartheta - 1) \left[M^2 - \tfrac{1}{3} S (S + 1) \right] + A \, M m, \tag{1.18}$$

resulting in 36 energy levels which are shown in fig. 1.2 for $\vartheta = 0^\circ$. In the absence of the static field and HFS, we have three nonequidistant energy levels (zero field splitting), which are each twofold degenerate due to Kramer's theorem. In the magnetic field, each level is split into two, giving six fine structure levels. HFS causes a further splitting. It is to be noted that the levels do not vary linearly with increasing field if one considers higher order terms.

For the ESR transitions we find

$$h\nu = E_{M,m} - E_{M-1,m} = g\mu_{\mathrm{B}} H_{M,m} + \tfrac{1}{2} D (2M - 1) (3 \cos^2 \vartheta - 1) + A m. \tag{1.19}$$

Dividing by $g\mu_{\mathrm{B}}$ and defining $H_0 = h\nu/g\mu_{\mathrm{B}}$, i.e. H_0 is the centre of the spectrum, we get for the positions $H_{M,m}$ of the various lines in terms of the magnetic field

$$H_{M,m} = H_0 - \tfrac{1}{2} D' (2M - 1) (3 \cos^2 \vartheta - 1) - A' m. \tag{1.20}$$

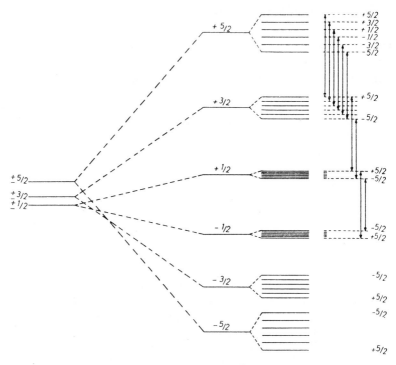

Fig. 1.2. A schematic energy level diagram for the Mn^{2+} ion (S = 5/2; I = 5/2), showing the electronic levels in zero and in strong magnetic field, together with the splitting due to the nuclear spin. Some typical transitions are indicated by arrows.

Now D' and A' are measured in G. For $A' < D'$, the spectrum consists of $2S$ (i.e. five) fine structure groups, each with $2I + 1$ (i.e. six) HFS components, the spectrum being symmetric about H_0 and the separation of the groups depending on the angle, according to the term proportional to $3 \cos^2 \vartheta - 1$. This behaviour can be seen in fig. 1.3.

Fine structure transitions in general are of unequal intensity, the intensity varying as the product

$$\delta N \left[S (S + 1) - M (M - 1) \right], \tag{1.21}$$

where δN is the population excess in the lower state of each transition. For thermal equilibrium and high temperatures δN is constant, and in the case $S = \frac{5}{2}$ the relative intensities of the five FS components are 5 : 8 : 9 : 8 : 5.

This simplified picture is modified by the other terms in the Hamiltonian (1.17) as well as by higher order corrections of the perturbation treat-

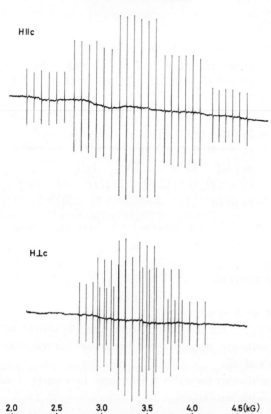

Fig. 1.3. ESR of Mn^{2+} in a ZnO single crystal ($\nu = 9.5$ GHz, $T = 300$ K). The spectrum consists of five fine structure groups, each with six equally intense HFS lines. The spectrum is strongly angular dependent. Recorded is the first derivative of the ESR absorption.

ment. For small line widths, third order corrections in D and A have to be considered to determine the parameters of the Hamiltonian with the accuracy wanted. From third order perturbation treatment one obtains the following expressions for the line positions at the orientations $H\|c$:

$$
\begin{aligned}
H_{M,m} = H_0 &- D'\,(2M-1) - A'm \\
&+ \tfrac{7}{36}\,(a'-F')\,\{4\,M^3 - 6\,M^2 + \tfrac{2}{7}\,M\,[19 - 6\,S\,(S+1)]\} \\
&+ (\tfrac{1}{2}\,A'^2/H_0)\,[I\,(I+1) - m^2 + m\,(2\,M-1)] \\
&- (\tfrac{1}{2}\,A'^2 D/H_0^2)\,\{[3\,I\,(I+1) - 3\,m^2]\,(2\,M-1) \\
&\quad - m\,[2\,S\,(S+1) - 6\,M^2 + 6\,M - 3)]\} \\
&+ (\tfrac{1}{2}\,A'^3/H_0^2)\,\{[2\,I\,(I+1) - 3\,m^2]\,(2\,M-1) \\
&\quad - m\,[S\,(S+1) + I\,(I+1) - 3\,M^2 + 3\,M - m^2 - 2]\};
\end{aligned}
$$

$$(1.22)$$

and for $H \perp c$:

$$
\begin{aligned}
H_{M,m} = H_0 &+ \tfrac{1}{2} D' (2M - 1) - A'm \\
&+ (\tfrac{1}{8} D'^2/H_0) [6M^2 - 6M - 2S(S+1) + 3] \\
&+ (\tfrac{1}{16} D'^4/H_0^2) \{8M^3 - 12M^2 + 18M - 7MS(S+1) \\
&+ \tfrac{7}{2} S(S+1) - 7\} \\
&+ \tfrac{7}{96} (a' - F') \{4M^3 - 6M^2 + \tfrac{2}{7} M [19 - 6S(S+1) \\
&+ \tfrac{6}{7}(S(S+1) - 2]\} \\
&- (\tfrac{1}{2} A'^2 D/H_0^2) \{(2M - 1) [I(I+1) - m^2] \\
&+ m [M^2 - M + 1 - S(S+1)]\} \\
&- (\tfrac{1}{2} A'^3/H_0^2) \{2(2M - 1) [I(I+1) - m^2] \\
&- m [S(S+1) - 3M^2 + 3M - 2]\}.
\end{aligned} \tag{1.23}
$$

In eqs. (1.22) and (1.23) the various terms have been summed due to their relative order of magnitude.

1.5. Hyperfine structure

1.5.1. HFS OF AN S STATE ION

At this point of the discussion, some remarks should be made on the HFS of an S state ion, because a half filled shell for transition ions does not show any HFS at all.

One often divides the HFS tensor into two parts, a scalar part and a tensor part with zero trace:

$$
A_{ij} = a + B_{ij}, \qquad \sum_k B_{kk} = 0. \tag{1.24}
$$

Here a denotes the isotropic 'contact' term (the so-called 'Fermi contact term') which is related to the electron density at the site of the nucleus by

$$
a = \tfrac{8}{3} \pi \, \mu_B g_s g_I |\psi(0)|^2. \tag{1.25}
$$

B is the tensor indicating the anisotropic HFS interactions. It can be understood in terms of dipole–dipole interaction of the electrons and nuclear spins. The isotropic part is in first approximation related to unpaired s electrons, and the anisotropic residue gives information about the spatial distribution of the electron spins.

For the simple case of axial symmetry, that means the nucleus lies on an axis of > second order symmetry of the electron distribution, the B tensor has the following form

$$B_{xx} = B_{yy} = -b, \qquad B_{zz} = 2b, \tag{1.26}$$

and the HFS interaction energy is

$$E_{\text{HFS}} = a + b \, (3 \cos^2 \vartheta - 1). \tag{1.27}$$

For Mn^{2+} in ZnO, however, a is much greater than b, so that b can be neglected.

To understand the rather large HFS splitting which is observed for the S state ions, it can be assumed that unpaired s electrons contribute to the HFS splitting via the Fermi contact term. Thus the orbital wave functions differ if the s electrons are with spin parallel or antiparallel to the d electrons because of an exchange term which is absent in the antiparallel case. Since the spins of the d electrons are all parallel for the $3d^5$ $^6S_{5/2}$ state, the spin dependent exchange interaction polarizes the s^2 core electrons. The polarized core electrons in turn produce a magnetic field at the site of the nucleus (Abragam et al., 1955; Heine, 1957; Wood and Pratt, 1957; Watson and Freeman, 1961).

1.5.2. HFS INTERACTION WITH SEVERAL NUCLEI

Generally, in considering defects in solids the electrons interact with many nuclei. This slightly changes those terms in the Hamiltonian that contain the nuclear spin operator I. They are then to be summed up over all interacting nuclei $1, \cdots, n$ with which the electron has a measurable interaction. The energy eigenvalues may to a first approximation be written as the sum of the interactions with these nuclei:

$$E = \text{electron Zeeman term} + \text{FS term} + \sum_{j=1}^{n} (A_j M m - g_{Ij} \mu_N H m_j). \tag{1.28}$$

Here g_{Ij} is the nuclear g factor and μ_N the nuclear magneton.

Equivalent or nearly equivalent nuclei can be collected into groups with a total quantum number

$$M_I = \sum_{i=1}^{K} m_i = KI, \ KI - 1, \cdots, -KI.$$

Depending on the number of possible combinations of the m_i to form M_I, the individual values of M_I have different statistical weight. For instance, for $I = \frac{1}{2}$ and three equivalent nuclei the statistical weights are 1, 3, 3 and 1 for $M_I = \frac{3}{2}, \frac{1}{2}, -\frac{1}{2}$ and $-\frac{3}{2}$, respectively. The hyperfine splitting of the Zeeman levels produced by equally coupled protons is such a case and is shown in fig. 1.4 for two and three equivalent nuclei with nuclear spin $I = \frac{1}{2}$ interacting with an unpaired electron. A further example is given in section 2.2.

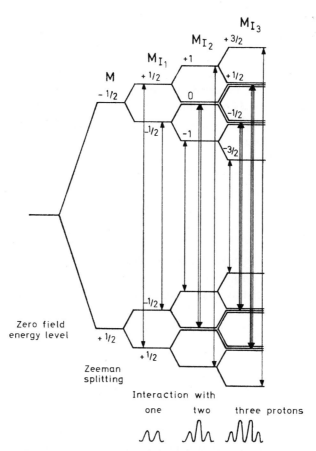

Fig. 1.4. HFS splitting of the Zeeman levels produced by equivalent protons interacting with an unpaired electron.

1.6. Determination of the parameters. Forbidden transitions

If the spectrum is well resolved, one can determine S and I by counting fine structure and hyperfine structure components, respectively. The electronic g factor can be determined by measuring the microwave frequency and the magnetic field corresponding to the resonance transitions. There are two main difficulties in the experimental determination of g:

(i) If the lines we are dealing with are rather broad, let us say of some 50 G halfwidth, the uncertainty in the accuracy of the g value lies in the determination of the centre of the line.

(ii) Another complication occurs if the magnetic hyperfine interaction is not negligible in comparison to the Zeeman energy. In this case, second order corrections to the resonance condition have to be considered.

From eqs. (1.22) and (1.23), the parameters D, $|a - F|$ and A can be calculated as well as their relative sign. To first order, neither quadrupole nor nuclear Zeeman interactions effect transitions for which $\Delta m = 0$. Often, however, one observes so-called 'forbidden' transitions having considerable intensity. They occur if the quadrupole interaction tries to align the nucleus along the symmetry axis of the crystalline field whereas the magnetic field established by the electrons tries to align the nuclear spin at right angles to it. Thus nuclear states become involved, giving rise to off-diagonal terms of the form $(S_z S_+)(S_- I_+)$ and $(S_z S_-)(S_+ I_-)$ and the normal selection rules break down. Here

$$S_+ = S_x + iS_y, \qquad S_- = S_x - iS_y. \tag{1.30}$$

The line positions for the forbidden transitions can approximately be calculated and can be used for determining the nuclear g factor and the parameter P. Also for other strongly anisotropic interactions one must take into account the fact that the nuclear spin is no longer quantized in the direction of H, but rather in the direction of an effective field H_{eff}, which is the vector resultant of the external field and a fictitious field arising from the electron–nuclear interaction.

If the angle between H_{eff} and H gets large, nuclear states become admixed and forbidden transitions show up. Spectra of forbidden transitions are shown in figs. 1.5 and 1.6. These transitions may have appreciable intensity, becoming even more intense than the allowed transitions. However, the overall intensity remains constant due to the fixed number of centres. For the details of such considerations see for instance Slichter (1963) or Hausmann and Huppertz (1968).

A fact that often complicates the spectra to a further extent is the following: If the centre is anisotropic and oriented in a particular direction of the crystal lattice, there are normally several crystallographically equivalent orientations of the centre. Centres are to be found in all orientations with equal probability. As these centres are not equivalent with reference to H, one observes lines at different fields for each orientation of the crystal in the magnetic field H. At special orientations, the lines partially collaps

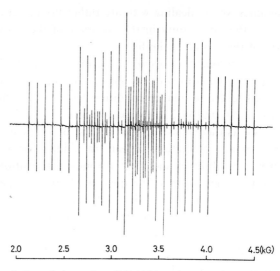

Fig. 1.5. Groups of allowed $\Delta m = 0$ and forbidden $\Delta m = 1$, $\Delta m = 2$ transitions in the spectrum of ZnO : Mn^{2+} ($\vartheta \approx 11°$, $\nu = 9.5$ GHz, $T = 300$ K).

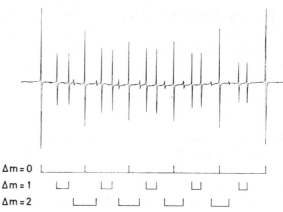

Fig. 1.6. Forbidden transitions of ZnO : Mn^{2+} at intermediate angle. ($\vartheta \approx 10°$, $\nu = 9.5$ GHz, $T = 300$ K). The overall intensity of the spectrum remains constant. These lines belong to the central group of fig. 1.5.

since some crystallographic axes may have the same angle with the field. For example, if $H\|[100]$, the four body diagonals of the cubic lattice have an angle of $\approx 54°$ with [100]. Thus the four lines of Se$^-$ centres in KCl which are oriented along the body diagonals occur at the same magnetic field.

The angular variation of the ESR transitions of O$^-$ centres in KCl is

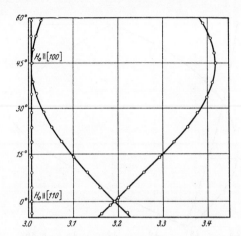

Fig. 1.7 (a). Angular variation of the ESR spectrum of O⁻ ions in KCl. As the ions are oriented along the three ⟨100⟩ axes with equal probability, there are three equally intense lines for arbitrary orientation of the magnetic field. For $H \parallel$ [100] and $H \parallel$ [110] the transitions partially collaps.

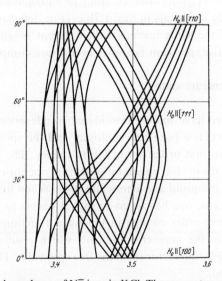

Fig. 1.7 (b). Angular dependence of N₂⁻ ions in KCl. The symmetry axis of the ions is along the face diagonals. Therefore there are six groups of lines each of which is split into five HFS lines. For special orientations of the magnetic field the transitions occur at the same field.

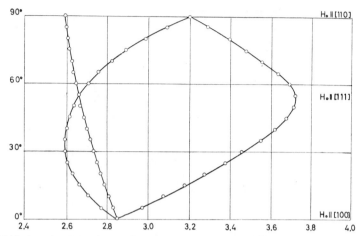

Fig. 1.7 (c). Angular dependence of Se⁻ ions in KCl. The orientation of the centres is along the four body diagonals, resulting in four equally intense signals. At $H \parallel$ [100] the lines collaps as all body diagonals have the same angle with H.

shown in fig. 1.7a. The ions are oriented along $\langle 100 \rangle$ axes and the spectrum consists of lines corresponding to centres along the three cubic axes. Similar arguments hold for N_2^- ions oriented along face diagonals, and for Se⁻ ions oriented along body diagonals in figs. 1.7b,c. (For details see section 4.4.)

If isotopes of the same nucleus with different magnetic moments are involved in HFS, the spectrum becomes even more complicated.

1.7. Orbitally degenerate states

Up to now, we have mainly considered nondegenerate states. In this case, the lowest level is a pure spin multiplet and the spin–orbit interaction gives no splitting in first order perturbation theory. The reason is that the matrix elements of the orbital angular momentum operators for each electron within the ground manifold are identically zero. When first order perturbation theory gives zero, one has to go to higher orders to see what splitting may result. The first order change in the wave function gives rise to first order corrections in the magnetic moment, which result in a deviation of the spectroscopic splitting factor from the free spin value. Hence the magnetic moment is modified.

When the crystal field leaves the lowest state orbitally degenerate, there may be nonvanishing matrix elements of the orbital angular momenta, and hence of the spin–orbit interaction. The detailed treatment of the spin–orbit contribution in an arbitrary symmetry requires a rather complicated calculation. As a result, one finds that the g tensor becomes strongly anisotropic even with moderate deviations from a high symmetry as the octahedral one. This behaviour is found for most of the transition row elements. The effect is for instance very clearly seen for divalent cobalt ions in ZnO crystals. Here both the g and A tensors are highly anisotropic, g_{ij} varying from 2.25 to 4.55 and A_{ij} from 15.3 to 2.8 G. Fig. 1.8 shows the spectrum of Co^{2+} in ZnO which consists of eight HFS lines of equal intensity due to the interaction with the Co^{59} nucleus.

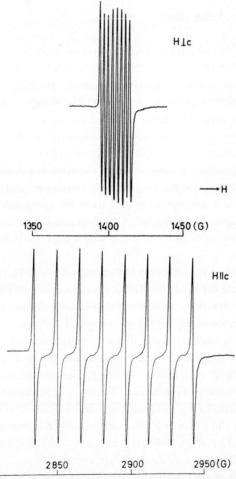

Fig. 1.8. Spectrum of Co^{2+} in ZnO ($T = 10$ K). The spectrum reveals the strong anisotropy of the g tensor and the A tensor. Recorded is the derivative of the absorption.

Here we should mention the sensitivity of the magnetic moment to small distortions in case the moment contains strong orbital contributions, which in turn are sensitive to the crystal field. For an S state ion this effect is only very small.

Orbitally degenerate centres are also much stronger affected by the thermal fluctuations of the environment than an S state ion, giving rise to rapid relaxation which broadens the ESR at room temperature beyond detection. For a more detailed discussion of these points the reader is referred to Low (1960) and to the literature cited therein.

1.8. The Jahn–Teller effect

A discussion of paramagnetic centres is incomplete without mentioning the Jahn–Teller effect. The effect occurs when a system of electrons and nuclei has degenerate or nearly degenerate electronic energy levels (non-Kramer's degeneracy) that are reasonably sensitive to distortions of the nuclear framework. The Jahn–Teller effect may simply affect the values of the parameters entering the spin Hamiltonian of the centre, causing these values to be anomalous from the standpoint of crystal field theory, or it may appear as a spontaneous distortion of the complex towards lower symmetry than that of the crystalline environment in the undisturbed symmetry of the host lattice. For a degenerate lowest state, the symmetrical surroundings are unstable. Suitable distortion splits the degeneracy and lowers the energy of at least one of the levels by an amount which depends linearly on the extent of the distortion.

The fundamental theorem of the Jahn–Teller effect was proved in 1937. A review article on the Jahn–Teller effect in solids has been given by Sturge (1967), where detailed calculations can also be found. Let E_0 be the energy of the degenerate state for the symmetrical configuration. Then according to the theorem there is some distortion Q that splits the degeneracy and causes the lower split-off state to have an energy E_-, such that $E_0 - E_-$ increases linearly with Q, for small Q. The quasielastic energy tending to keep the environment symmetrical increases quadratically. Hence the system can lower its energy by distorting to a finite Q. A new stable equilibrium is reached when the distorting force on the lattice is balanced by the restoring elastic force. The energy difference E_{JT} by which E_- lies below E_0 is the energy of stabilisation or Jahn–Teller energy.

It is evident that thus the original symmetrical situation is unstable. However, there may of course be more than one position of stable equilibrium, corresponding to positive and negative values of Q as well as to other modes of distortion. If the energy gained by distortion is small compared with the zero point energy of the associated lattice vibrations, we have a weak Jahn–Teller effect and the magnetic behaviour is the average of that in the equivalent distorted positions. That means, the Hamiltonian has still the same symmetrical form as before though the parameters are modified. In this case distortion does not actually remove the degeneracy as the system can resonate between the equivalent positions.

In an ESR experiment at low temperatures, so that thermally activated reorientation can be ignored, one would expect that the system spontaneously distorts to one of the positions of stable equilibrium. As there are several energetically equivalent distortions, one would expect the centres in a crystal to choose among these. The spectrum would consist of the spectra of the different distortions. However, this is only valid if the coupling between the distortions and the electrons of the centre is strong. Then the vibrational modes of motion can no longer be ignored. We have the so-called dynamic Jahn–Teller effect. Here the distorted configurations minimizing the energy are so different from one another that resonance is infinitely slow. When tunneling is also slow compared with the paramagnetic resonance frequency, the distorted configurations can be regarded as being static. Now the centre may be viewed as being basically asymmetric, the degeneracy being lifted. The change in energy by distortion can be calculated by the same crystal field methods as have been described earlier.

As a result of the strong Jahn–Teller effect, the environment in many cubic compounds is a distorted octahedron with four ligands close together in a square, the other two being much farther apart.

Similar arguments as for the Jahn–Teller effect also hold for other external perturbations, in particular strain and applied electric fields. Here the spin Hamiltonian of the defect is modified by these perturbations.

1.9. Line width and relaxation times

Often in spin resonance experiments the resolution of fine structure and hyperfine structure is limited by the line width of the absorption lines. This line width is influenced by several effects such as the interaction between the paramagnetic ions and the surroundings (spin–lattice relaxation), the interaction between the paramagnetic ions themselves (spin–spin relaxation) and by inhomogeneities in the crystal lattice and the applied external field. Further reasons for broadening are exchange interactions, saturation by the radiation field and unresolved hyperfine structure interaction.

In 1946, Bloch proposed a set of phenomenological equations to describe the dynamic magnetic behaviour of interacting nuclear paramagnets. These equations introduce the relaxation times T_1 and T_2, which for electrons have different magnitudes from those Bloch (1946) used to describe nuclei. However, the underlying physical considerations apply to both electrons and nuclei.

Our experiments observe a macroscopic property of the sample, conveniently the magnetization M, which is defined as the magnetic moment per unit volume. Bloch supposed that the interactions to be considered cause the magnetization to tend exponentially towards its thermal equilibrium magnitude $M_0 = \chi_0 H$, once the equilibrium has been disturbed. For $\mu_B H/kT \ll 1$, χ_0 is the static Curie susceptibility,

$$\chi_0 = \frac{Ng^2\mu_B{}^2 J(J+1)}{3\,kT}. \tag{1.32}$$

Here N is the number of magnetic moments and J the total angular momentum. For $N = 10^{22}$, at room temperature $\chi_0 \approx 10^{-5}$.

We may imagine that somehow a nonequilibrium magnitude and orientation of M has been achieved, let us say by pulsing the H_{rf} field and applying it to a sample initially having $M = M_0$. Turning off the H_{rf} field after a short period in general leaves M pointing in a nonequilibrium direction although it still has its initial magnitude M_0. When the nonequilibrium situation is achieved, M precesses at a frequency $\nu = g\mu_B H/h$ about H except for the effects of interaction between the spins, i.e. effects of relaxation.

As the individual moments μ_i making up M flip to bring the measurable component of the μ_i along the direction of H, which is M_z, towards the equilibrium value M_0, the z component of M will grow. That means that the moments transfer energy to the lattice. This process is called spin–lattice relaxation and is described by the spin–lattice or longitudinal relaxation time T_1. This same process also causes the component $M_\perp$ normal to M_z to decay.

Yet we have to consider another process not requiring energy exchange with the lattice. The individual moments μ_i each see a local dipolar field of their neighbours which adds vectorially to H; sometimes subtracting from, sometimes adding to H. Thus each μ_i in general precesses in a slightly different field and the coherence in the individual components making up the component $M_\perp$ is lost, causing $M_\perp$ to decay. This process is described by the spin–spin relaxation time T_2.

1.9.1. Spin–lattice relaxation

The study of spin–lattice relaxation in solids has a longer history than magnetic resonance. The early experiments for achieving low temperatures by adiabatic demagnetization have put attention on the coupling between spin and lattice degrees of freedom. The theory of spin–lattice relaxation has been started by Waller (1932). We can think of the mechanisms as follows:

When the lattice vibrates, the interionic distances are modulated at the frequency of the lattice vibration. An oscillatory magnetic field component thus arises from the motion of the neighbouring magnetic dipoles, and those lattice vibrations with frequency $v = g\mu_B H_0/h$ have a nonzero probability for flipping a moment from 'up' to 'down' accompanied by the emission or creation of a phonon having $hv = g\mu_B H_0$. This single phonon process is called the 'direct process'.

A second process, called 'indirect', is the one in which a higher frequency phonon is inelastically scattered by the flipping spin. The indirect or Raman process is of higher order than the direct one. Only at high enough temperatures, however, the indirect process has some importance as it utilizes all the modes of the spectrum and thus brings into play the energetic and far more plentiful higher frequency modes near the cut-off of the acoustic vibration spectrum.

In recent years, many difficulties have appeared which cannot be accounted for by a simple direct spin–phonon interaction. Often one finds relaxation times by a factor ten larger than they should be. Mostly it is still impossible to predict quantitatively what relaxation time a given ion will have or how it is influenced by crystal field strength or crystal symmetry. Possibly, a process has to be taken into account which Van Vleck (1939) already considered: He discusses a mechanism in which the lattice vibrations can influence T_1 by modulating the crystalline electric field, which in turn is felt by the spins through spin–orbit coupling.

1.9.2. Spin–spin interaction

Dipolar broadening is caused by two interactions. First the various ions set up magnetic fields at any other ion. These ions then see a magnetic field different from the external field. In the case of paramagnetic defects, the effect is generally small but for concentrated paramagnetic salts it may be of considerable magnitude. A nearest neighbour ion having a magnetic moment of 1 μ_B at a distance of 3 Å produces a magnetic field of about 300 G; the sum for all nearest neighbours may well exceed 500 G.

The second mechanism causing dipolar broadening only operates between like ions. Two ions precessing with the same Lamor frequency interact and cause their spin orientation to change. This reduces the average lifetimes of the ion in a given state, thus producing broadening of the lines.

1.9.3. EXCHANGE INTERACTION

In crystals with high concentrations of paramagnetic ions, one only rarely finds agreement between the experimental and theoretical line width. Dirac (1935) showed that one of the main reasons for this is an electrostatic exchange effect connected with the overlap of the orbital wave functions of the electrons. The exchange coupling integral decreases rapidly with distance and becomes negligible in dilute crystals, but may play an important role in concentrated crystals. If only interaction between nearest neighbours is considered, the interaction is of an isotropic form, whereas dipolar broadening is not spatially invariant. However, even with these simple assumptions the measurements of line width and shape and their interpretation are rather difficult.

1.9.4. OTHER REASONS FOR BROADENING

In experiment, one often finds that there is a residual line width which cannot be attributed to spin–spin or spin–lattice relaxation or to exchange coupling mechanisms. This additional line width is particularly great for cases with large zero-field splitting and depends on the quality of the single crystal samples. It is thought that the line width may be caused by small variations in zero-field splittings. These may be produced by defects and dislocations or twinning and mosaic structure in the vicinity of the magnetic ions. Inhomogeneities of the applied magnetic field may give rise to similar broadening effects.

1.9.5. MEASUREMENTS OF RELAXATION TIMES

The relaxation time T_1 is normally measured by saturation or by pulse methods. The saturation method may be simply understood by noting that if a spin system is made to absorb power, then, in steady state, this energy must at the same rate be passed on to the lattice that forms the thermal reservoir. This rate of energy flow per unit volume can be expressed by the spin level population densities and is set equal to the power absorbed by the spin system per unit volume. From this it follows that by increasing the field H_{rf} until the power absorption becomes nonlinear with input power one can determine the value of T_1 (see section 4.1). The spin–spin relaxation time can be obtained from careful measurements of the line shape at very great dilution of the magnetic ions.

The most reliable way to measure the relaxation times T_1 and T_2 is the spin-echo method (Hahn, 1950). This method gives the most accurate and direct measurements of T_1; however, the difficulty is that one has to provide very short pulses for substances having short relaxation times, and fast pulse techniques in the microwave range are not easy.

New results have been achieved for the measurements of the relaxation time T_1 at low temperatures by utilizing the magneto-optical properties of centres, i.e. circular dichroism and Faraday rotation (Karlov et al. 1963; Mort et al., 1965).

1.10. The information obtained by ESR spectra

The information given by an ESR spectrum can be classified under several subheadings.

(i) If ESR spectra are detected, one can decide that a centre is paramagnetic. The other way round cannot be concluded as the detection strongly depends on the relaxation times. If there are doubts in identifying paramagnetism or diamagnetism, measurements of the static susceptibility should be made. However, this may not always be possible if one is dealing with low concentrations of defects.

(ii) The area under the absorption curve is related to the number of spins. Therefore the determination of the concentration of centres and their absolute number is possible. This fact may sometimes be utilized for measuring the purity of crystals. It has for instance been shown by means of ESR measurements that the content of OH^- ions in alkali halides can be determined (Gründig and Wassermann, 1963). Often optical absorption measurements can also be related to ESR spectra and thus the oscillator strength in the optical bands can be determined with greater accuracy.

(iii) The position of the resonance on the recorder trace gives the field of which resonance occurs, and from this the g value can be calculated. The g factor is a measure for the effective moment of the dipole. The deviation from the free electron value tells us about the amount of orbital states involved in the ground state. Even though for many defects, especially for colour centres, the g shifts and its anisotropy are only small because we are dealing with S type ground states or because of the shielding of the electron orbital momentum by the crystalline field, important information can be evaluated about the structure of the centre.

(iv) The variation of the resonance fields with the angle between external field and symmetry axis reflects the local symmetry. That is, it may be decided, whether the symmetry is cubic, axial or lower. The spectra provide a powerful method of determining small deviations from the overall symmetry; deviations which are so small that they might not be detected by X-rays, but result in an appreciable fine structure splitting. The occurrence of fine structure splitting, which can be identified by the angular dependence of the spectrum, has for instance given a definite confirmation of the aggregate models of the M centre and the R centre in the alkali halides (see chapter 4). Both are F aggregate centres with excited states having total spin $S > \frac{1}{2}$.

(v) The most important information may be obtained from HFS. The number of lines and their relative intensities give information about the sort and the number of nuclei involved in the defect. The amount of the HFS splitting is related to the electron density at the nucleus, the anisotropy of the HFS informs about the anisotropy of the electron density and also reflects the local symmetry of the centre.

It should be noted, however, that the influence of neighbouring nuclei, giving rise to unresolved inhomogeneous broadening of the lines, is often rather disturbing. Here ENDOR may bring help. Information about the quadrupole moment is also obtained from the HFS splittings. Another point is that the variation of the HFS with covalency for a certain ion in different hosts may give information about the ionicity of compounds.

(vi) Reversible and irreversible dynamic processes concerning excitation and transformation as well as production and diffusion of defects may be followed in the ESR spectra. Quite often the activation energy of certain processes may be obtained from the temperature dependence of the spectra.

(vii) The spin–lattice relaxation time T_1 measures the energy transfer from the paramagnetic centre to the surrounding, that is to the lattice. Thus we gain information about the strength and the nature of the coupling of the spin system to the lattice. Via 'direct interaction' of the spins with the lattice we may find out something about the energy distribution of the phonons at low temperature. Measurements of T_2 say something about the coupling of the electron spins with one another.

This information is obtained from investigations of the width of a resonance line and the saturation behaviour. The mechanisms which broaden lines also provide information about the resonant site. It is convenient to divide the mechanisms in those causing homogeneous and inhomogeneous broadening. Examples for the first are (1) spin–lattice relaxation, (2) dipolar broadening between like spins and (3) interaction of the radiation field with the spins. Examples for inhomogeneous broadening are (1) unresolved HFS interaction and (2) inhomogeneities in the crystal field (small angle grain bounderies).

(viii) The line shape function represents the shape of the absorption line. For nearly all samples it approximates to a Gaussian or Lorentzian distribution, depending on which type of interaction is the main source of broadening. The line shape is often very important in determining the correct g value of a resonance and is very much influenced by effects such as saturation broadening, exchange narrowing or modulation broadening. Knowledge about these mechanisms can be evaluated from the line shape.

1.11. Experimental techniques

1.11.1. GENERAL

In its basic items an ESR spectrometer works like an optical absorption spectrometer. The monochromatic radiation of a klystron is passed through a waveguide to a detector (fig. 1.9). The sample containing unpaired electrons is placed into the waveguide and an external magnetic field is applied. If, when slowly increasing this field, the resonance condition $hv = E = g\mu_B H$ is fulfilled, absorption and induced emission processes take place between the electron spin levels in the sample. Because the lower levels are more

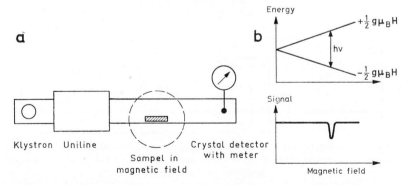

Fig. 1.9. (a). Simple arrangement for measuring ESR absorption. (b). Energy levels and detector signal with increasing magnetic field.

populated, a net absorption is observed at the detector. Such a simple spectrometer has a rather low sensitivity, but can be made to show the resonance effect by using a 'strong' sample (1 g of the radical diphenylpicrylhydrazyl (DPPH)).

Numerous efforts have been made to increase the sensitivity. These include utilization of (1) resonant cavities with high quality factors Q as absorption cells, (2) field modulation and a.c. detection, (3) narrow band technique with a phase sensitive detector and (4) superheterodyne detection. Of course, stabilization of the microwave frequency and of the magnetic field are of great importance with these advanced methods.

These detection techniques have been reviewed recently by Ingram (1970) and need not be discussed in detail here. Commercially available ESR spectrometers now work at the X band (microwave frequency about 9 GHz, wavelength about 3 cm, magnetic field about 3000 G) or at the Q band (35 GHz, 8 mm and 13000 G, respectively). They most commonly use 100 kHz field modulation (fig. 1.10) and are able to detect 10^{10} spins (X band) or 10^9 spins (Q band) if the signal line width is 1 G.* Superheterodyne attachments are

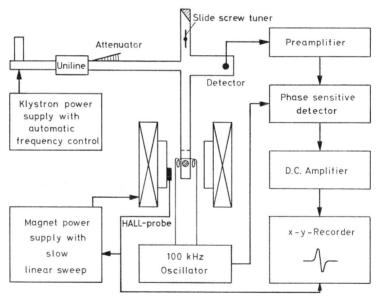

Fig. 1.10. Block diagram of a modern ESR spectrometer with 100 kHz field modulation, phase sensitive detection. The spectrometer records the first derivative of the absorption line with an XY recorder.

* A sensitivity several orders of magnitude better may be achieved in combination with optical techniques (see section 4.1).

available too. They have slightly better sensitivity and are used in combination with the ENDOR technique (see chapter 2).

The better sensitivity of the Q band systems results from the increased population difference between adjacent spin levels at the higher magnetic field. They have the further advantage that the spectra can more easily be analysed at higher magnetic fields, because terms of higher order in calculating the energy levels (see section 1.4) are of less importance.

The resolution of modern ESR spectrometers is well better than 100 mG. This is sufficient for the study of defects in solids since the observed lines are mostly inhomogeneously broadened. The reported line widths range from about 30 mG to well above 1000 G.

1.11.2. SPECIAL TECHNIQUES

For defect studies, some auxiliary techniques may be necessary. These involve applying uniaxial or uniform stress to the sample or subjecting it to irradiation with light, X-rays or fast particles.

1.11.2.1. Application of low temperatures.   This is often necessary since some defects are unstable at higher temperatures, and intensities and line widths of the ESR lines are often much more favourable at low temperatures. Some types of cryostats shall be discussed now.

1.11.2.2. Normal cryostats.   A single or double dewar system may be used with liquid nitrogen in the outer and liquid helium (or hydrogen) in the inner dewar. The cavity with the sample within is immersed into the helium bath. Heat input down the waveguide is prevented by installing a waveguide of thin-walled stainless steel or by a suitable interruption of about 1 mm between the high and the low temperature parts of the waveguide system. Commercially available He attachments of this type use cavities which are only suitable for audio frequency modulation. For high sensitivity requirements, superheterodyne detection has then to be used.

Several low temperature cavities have been constructed for use with the sensitive high frequency (100 kHz modulation) technique. As the cavity with the sample is immersed into the cooling liquid, the bubbling of the coolant causes some additional noise. This is most serious with liquid nitrogen and almost of no importance with liquid helium.

1.11.2.3. Cold finger cryostats.   These cryostats are used with the normal room temperature cavities, which are constructed for audio and high frequency modulation. Only the sample is cooled by immersing it into the cooling liquid (fig. 1.11). Vacuum isolation is provided by a quartz tube. The immersing system (fig. 11.1a) is very easy to handle, the sample can be rotated or rapidly changed using a suitable sample holder. This system can be used with liquid nitrogen, hydrogen or helium. In the latter case, an additional nitrogen shielding of the helium bath above the cavity is useful. The noise from bubbles in the liquid has already been mentioned.

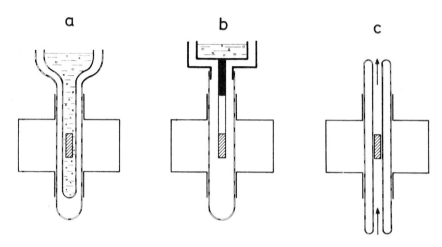

Fig. 1.11. Cold finger (a,b) and gas flow (c) cryostats for use with room temperature cavity resonators.

1.11.2.4. Gas flow cryostats.   While normal and cold finger cryostats are restricted in their temperature range to the temperatures accessible by the used coolants, the gas flow systems (fig. 1.11c) cover a wide temperature range and allow controlling or rapidly changing the sample temperature. Ordinary room temperature cavities are used. Systems for cooling down to nitrogen temperature are commercially available. The gas stream is first cooled by a nitrogen bath and then passed through the cavity by a double-walled, evacuated quartz tube. Controls are provided for the gas velocity and for its temperature by introducing a heating coil into the quartz tube outside the cavity.

This principle can also be used with hydrogen gas from a supply of liquid hydrogen. Temperatures down to 22 K may be achieved. A helium gas flow cryostat has been constructed by the authors, which uses the evaporation principle developed by Klipping (1961). This cryostat (fig. 1.12) cools the sample down to any temperature above 5 K within a few minutes

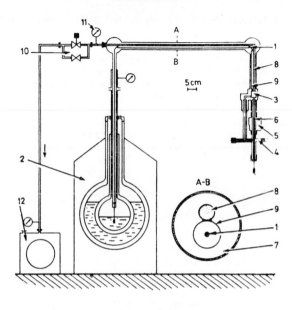

1 Pipe transporting liquid He
2 He storage vessel
3 Evaporator
4 Quartz glass system
5 TE 102 cavity
6 Sample
7 Vacuum system
8 Pipe transporting cold He gas
9 Radiation shield
10 Solenoid valve with by-pass
11 Pressure gauges
12 Feed pump

Fig. 1.12. Gas flow cryostat for the temperature range 5–300 K.

and allows rotating or rapidly changing the sample. The helium losses are cut down by using the cold gas leaving the cavity for shielding the incoming stream.

References

Abragam, A. and H. M. L. Pryce, 1951, Proc. Roy. Soc. London A **205**, 135.
Abragam, A., J. Horrowitz and M. H. L. Pryce, 1955, Proc. Roy. Soc. London A **230**, 169.
Bloch, F., 1946, Phys. Rev. **70**, 460.
Dirac, P. A. M., 1935, The principles of quantum mechanics (Oxford Univ. Press, London).
Gründig, H. and E. Wassermann, 1963, Z. Physik **176**, 293.
Hahn, E. L., 1950, Phys. Rev. **80**, 580.
Hausmann, A. and H. Huppertz, 1968, J. Phys. Chem. Solids **29**, 1369.
Heine, V., 1957, Phys. Rev. **107**, 1002.
Ingram, D. J. E., 1970, in: Handbuch der Physik Vol. **18** (1) (Springer, Berlin).

Karlov, N. V., J. Margerie and Y. d'Aubigué, 1963, J. Phys. (Paris) **24**, 717.
Klipping, G., 1961, Kältetechnik **13**, 250.
Low, W., 1960, Solid State Phys. Suppl. **2**.
Mort, J., F. Lüty and F. C. Brown, 1965, Phys. Rev. **137**, A 566.
Pryce, H. M. L., 1950, Proc. Phys. Soc. London A **63**, 25.
Slichter, C. P., 1963, Principles of magnetic resonance (Harper and Row, New York).
Sturge, M. D., 1967, Solid State Phys. **20**.
Van Vleck, J. H., 1939, J. Chem. Phys. **7**, 72.
Waller, T., 1932, Z. Physik **79**, 370.
Watson, R. E. and A. J. Freeman, 1961, Phys. Rev. **123**, 2027.
Wood, J. H. and G. W. Pratt Jr., 1957, Phys. Rev. **107**, 995.

2 | ELECTRON NUCLEAR DOUBLE RESONANCE

2.1. General

The most important informations about the nature and the structure of defects in solids are obtained by analysing the hyperfine structure (HFS) in spin resonance experiments. This HFS can be caused by the nucleus (or the nuclei) of the defect centre itself or by the surrounding nuclei, provided that the nuclei in question have magnetic moments (and perhaps electric quadrupole moments too). While the HFS interaction within the defect centre itself can easily be resolved by ordinary ESR (as was shown with Mn^{2+} in ZnO in section 1.4), the interaction with neighbouring nuclei contributes to the line width ('inhomogeneously broadened' line) and only in favourable cases can be resolved directly ('superhyperfine structure').

The ENDOR method provides a resolution several orders of magnitude better than the ordinary ESR. Its principle is explained now for a simple system: a defect with $S = \frac{1}{2}$ and $I = \frac{1}{2}$ (for instance H atoms in solids, P donors in Si).

According to section 1.3, the spin Hamiltonian is

$$H = \sum_{i,j} [g_{ij}\mu_B H_i S_j + A_{ij} S_i I_j - g_I \mu_N H_i I_j]$$ (2.1)

and consists of the electron Zeeman term, the HFS interaction and the nuclear Zeeman term. First order perturbation theory gives the energy levels

$$E_{M,m} = g\mu_B H M + E_{HFS} M\,m - g_I \mu_N H\,m,$$ (2.2)

with $M, m = \pm \frac{1}{2}$. These levels are sketched in fig. 2.1 for $g_I > 0$ and $\frac{1}{2} E_{HFS} > g_I \mu_N H$. ESR transitions fulfill the selection rules

$$\Delta M = \pm 1, \qquad \Delta m = 0.$$

In our example, two transitions are allowed with

$$h\nu_e' = g\mu_B H + \tfrac{1}{2} E_{HFS},$$
$$h\nu_e'' = g\mu_B H - \tfrac{1}{2} E_{HFS}.$$
(2.3)

The difference of these ESR lines, if resolved, gives the HFS interaction energy E_{HFS}.

ENDOR transitions are induced between different nuclear levels belonging to the same electronic level. They fulfill the selection rules

$$\Delta M = 0, \qquad \Delta m = \pm 1.$$

Our system therefore has two ENDOR transitions

$$h\nu_n^- = \tfrac{1}{2} E_{HFS} + g_I \mu_N H \quad (\text{for } M = -\tfrac{1}{2}),$$
$$h\nu_n^+ = \tfrac{1}{2} E_{HFS} - g_I \mu_N H \quad (\text{for } M = +\tfrac{1}{2}).$$
(2.4)

For $S = \frac{1}{2}$ and HFS interactions with several nuclei, such a pair of ENDOR transitions is observed for each nucleus. The frequency difference between these pairs is the double NMR resonance frequency of the nucleus in the applied magnetic field and so can serve to identify the involved nucleus.* The HFS interaction energy E_{HFS} can be measured with an accuracy of 10–100 kHz. So fine details of this interaction can be detected such as small deviations from axial symmetry, second order HFS and quadrupole effects.

The experimental realization of this method can not employ the direct detection of these transitions because of lack of sensitivity. Instead, the effect of the ENDOR transitions on the ESR absorption is observed. This is only possible if the ESR transition can (at least partially) be saturated. In an ENDOR experiment, the ESR resonance condition is kept constant with high energy input to achieve saturation, so that the ESR absorption signal is low (for example with the transition $h\nu_e'$ in fig. 2.1). Then high frequency radiation from a tunable oscillator is additionally applied to the sample. When ENDOR transitions are induced (i.e. $h\nu_n^+$ in fig. 2.1), the populations of the levels involved are changed (in fig. 2.1 the population of the upper level is reduced). This results in a desaturation of the ESR transition and in

---

* For $g_I \mu_N H > \frac{1}{2} E_{HFS}$, two ENDOR transitions result, which are equally spaced on each side of the NMR frequency and whose frequency difference is given by E_{HFS}/h.

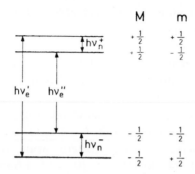

Fig. 2.1. Energy levels in a magnetic field for a system with $S = \frac{1}{2}$ and $I = \frac{1}{2}$ with ESR and ENDOR transitions ($g_I > 0$, $\frac{1}{2}E_{HFS} > g_{I}\mu_N H$).

an increase of the ESR absorption signal. Therefore this method has the name Electron Nuclear Double Resonance. As the effect is a small variation in the amplitude of the ESR signal, the sensitivity of ENDOR experiments is only some percent of that of ordinary ESR. Superheterodyne spectrometers are used and the defects have to be prepared in high concentrations.

The ENDOR method has been first introduced by Feher (1956, 1958). A stationary method, as described here, has been developed by Seidel (1961). A detailed treatment with several examples is given in the book by Abragam and Bleaney (1970). Different constructions were reported to solve the difficult problem of introducing the high frequency radiation (0.1–100 MHz) into the microwave cavity. Now, ENDOR attachments to ESR spectrometers are commercially available.

2.2. Example: The U_2 centre

The structure of the U_2 centre has been established by both ESR and ENDOR measurements. These centres are formed in alkali halide crystals doped with OH^- or SH^- ions by irradiation at low temperatures with ultraviolet light. According to the photochemical reaction

$$OH^- \xrightarrow{hv} O^- + H,$$

neutral hydrogen atoms ('U_2 centres') are formed. In the ultraviolet region, the optical absorption band due to OH^- is reduced and during this reaction two new bands appear which are ascribed to the new O^- and U_2 centres. U_2 centres in KCl are stable below 100 K.

Neutral hydrogen atoms are paramagnetic and can be detected by ESR. The ESR spectrum is split into two HFS components by interaction with the hydrogen nucleus ($I = \frac{1}{2}$). The HFS interactions of the U_2 centres with surrounding nuclei can also be resolved by ordinary ESR. Fig. 2.2 gives the ESR spectrum of U_2 centres in KCl, recorded at 20 K with H∥(100). The large doublet splitting (500 G) is due to the proton; the superhyperfine splitting ($\delta = 9$ G) is due to the nuclei of the neighbouring Cl⁻ ions.

Four equivalent Cl nuclei with $I = \frac{3}{2}$ have a total spin $I_{total} = 6$ and give the observed 13-line pattern with the intensity ratios $1 : 4 : 10 : 20 : 31 : 40 : 44 : 40 : 31 : 20 : 10 : 4 : 1$. From this, one can conclude that the H atom occupies a tetrahedral interstitial site in the KCl lattice (see fig. 2.3). By comparing these splittings in different host crystals (KCl, $\delta = 9$ G; KBr, $\delta = 48$ G; NaCl, $\delta = 16$ G), one sees that the interactions are due to the 4 halogen nuclei and not to the 4 alkali nuclei in the

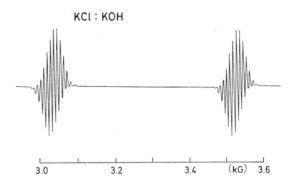

Fig. 2.2. ESR spectrum of the U_2 centre in KCl.

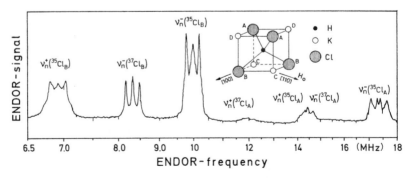

Fig. 2.3. ENDOR spectrum and model of the U_2 centre in KCl. (After Spaeth, 1966.)

same distance, which also have $I = \frac{3}{2}$. The interactions with the alkali nuclei are hidden in the ESR line width and therefore must be at least ten times smaller.

This model of the U_2 centre (hydrogen atom at an interstital site) also explains the observed angular dependence of the spectra. While for $H\|(110)$ a 19-line pattern and for $H\|(111)$ a 16-line pattern is found, for intermediate directions of the field the pattern is complicated and cannot be resolved by ordinary ESR.

Subsequent ENDOR measurements with U_2 centres have proved the given model beyond any doubt. It was possible to determine with high precision the superhyperfine interaction constants not only of the neighbouring halogen nuclei but also of the neighbouring alkali nuclei and even for the nuclei in the next nearest shells. As an example, the ENDOR spectrum for the Cl neighbour nuclei is given in fig. 2.3. There are Cl^{35} nuclei (abundance 74%) and Cl^{37} nuclei (abundance 26%), the ratio of their nuclear magnetic moments being $\mu_{37} : \mu_{35} = 0.832$. For $H\|(110)$, two pairs of equivalent Cl nuclei (A and B) are involved. The spectrum shows for each pair the ENDOR transitions $\nu_n{}^+$ and $\nu_n{}^-$. Due to quadrupole interaction, each transition is split into $2I = 3$ lines. A further splitting is caused by higher order effects. Intensities and frequencies of the Cl^{37} and the Cl^{35} lines correspond to the abundance and nuclear moment ratios.

The most striking result of the ENDOR work is the small interaction of the U_2 centre electron with the neighbouring K^+ nuclei compared to the strong, and detectable by ordinary ESR, interaction with the neighbouring Cl^- nuclei. This ratio is 1 : 24. Even if the lower magnetic moment of the K^+ nuclei is considered, the fact remains that the electron density at the Cl^- nuclei is 12 times higher than at the K^+ nuclei. Several attempts to establish a suitable U_2 centre wave function have so far not succeeded in explaining this problem.

References

Abragam, A. and B. Bleaney, 1970, Electron paramagnetic resonance of transition ions (Clarendon, Oxford).
Feher, G., 1956, Phys. Rev. **103**, 834.
Feher, G., 1958, Physica Suppl. **24**, 80.
Seidel, H., 1961, Z. Physik **165**, 218, 239.
Spaeth, J. M., 1966, Z. Physik **192**, 107.

3 | NUCLEAR MAGNETIC RESONANCE

3.1. General

The method of nuclear magnetic resonance (NMR) has been developed to measure gyromagnetic ratios of nuclei in liquids and solids. Numerous applications of this method deal with investigations concerning the structure and binding of inorganic and organic compounds. The so-called KNIGHT shift of the NMR signal in metals gives the possibility to determine conduction electron densities.

Applications of NMR to the study of defects are rather seldom as compared to ESR and ENDOR. The reason for this is the fact that the nuclear magnetic moments are three orders of magnitude smaller than the magnetic moment of the electron. Therefore, the NMR method is of considerably lower sensitivity than ESR. Hence, diluted impurities in solids cannot be studied directly by their own NMR signals.

In spite of this, defect studies by NMR are possible by investigating the resonance signals of the normal lattice nuclei. Their line widths and intensities may be changed by the presence of lattice defects. Mobile defects may cause a 'motional narrowing' of the lines, while fixed defects may cause line broadening or intensity decrease by the interaction of their additional electric field gradient with the quadrupole moments, if any, of the lattice nuclei. These field gradients can arise from lattice distortions near the defects or from additional charges in the case of impurities.

Results of such work have been compiled by Cohen and Reif (1957), in the book by Ebert and Seifert (1966) and by Kanert and Mehring (1970).

The principles of the NMR method and the interactions involved have been thoroughly treated in the book by Abragam (1961).

As an example, the influence of dislocations on the NMR signal is now discussed.

3.2. Example: Dislocations

In crystals with nuclear spins $I > \frac{1}{2}$, the interaction of the nuclear quadrupole moment Q with the electric field gradient at the nuclear site has to be considered. This gradient, and therefore the interaction, vanishes for nuclei in a perfect cubic lattice. Defects in such a lattice will cause distortions and by this field gradients in their neighbourhood. This effect is especially pronounced with dislocations. Moreover, their distortion fields vary as r^{-1}, as compared to r^{-3} with point defects. Therefore, dislocations are favourite defects for NMR studies.

Kanert (1964) investigated the influence of dislocations in plastically deformed alkali halide crystals. The energy levels in a magnetic field,

$$E_m = g_I \, \mu_N \, H \, m + E_Q(m),$$

are changed by a quadrupole interaction term $E_Q(m)$ which depends on the quantum number m. The corresponding line shifts and the remaining intensity in the central part of the undisturbed NMR line can be calculated from the distortion fields of the dislocations. The quadrupole interactions with nuclei in large distance from the dislocation core can be neglected, so these nuclei give their full contribution to the line intensity. In the medium range, the satellite transitions ($m \neq \frac{1}{2}$) are so much displaced that they give no intensity contribution to the central part of the line, while the principal transitions between the levels $m = \frac{1}{2}$ and $m = -\frac{1}{2}$, which are less influenced by quadrupole interaction, give their full contribution. The nuclei near the dislocation core do not contribute to the line intensity as all their transitions are displaced too much.

Fig. 3.1 gives the measured intensity of the Na^{23} resonance line in deformed NaCl crystals as a function of the dislocation density which was determined by the etch pit technique. According to theory, an intensity decrease of 60% is observed for dislocation densities greater than 10^7 cm^{-2}. Then, most lattice atoms are located in the above mentioned medium range. Similar experiments with NaF crystals again give an intensity decrease of the Na^{23} resonance while the F^{19} resonance remains unchanged, as the F^{19} nucleus has $I = -\frac{1}{2}$ and no quadrupole moment. The observed total line

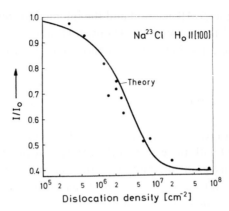

Fig. 3.1. Intensity of the Na^{23} NMR signal in plastically deformed NaCl crystals. (After Kanert, 1964.)

width, defined by the extremal values of the differentiated absorption curve, is nearly constant with increasing dislocation density and shows only a small maximum in the region of the strong intensity drop (Kanert, 1965).

The NMR experiments described here give the possibility to determine dislocation densities. They have some advantage over the etch pit technique as they give an average density over the whole crystal. The method has been successfully extended to dislocations in f.c.c. metals (Kanert, 1969).

References

Abragam, A., 1961, The principles of nuclear magnetism (Oxford Univ. Press, London).
Cohen, M. H. and F. Reif, 1957, Solid State Phys. **5**, 321.
Ebert, I. and G. Seifert, 1966, Kernresonanz in Festkörpern (Akademische Verlagsgesellschaft, Leipzig).
Kanert, O., 1964, Phys. Stat. Sol. **7**, 791.
Kanert, O., 1965, Z. Physik **184**, 92.
Kanert, O., 1969, Phys. Stat. Sol. **32**, 667.
Kanert, O. and M. Mehring, 1970, Static quadrupole effects in disordered cubic solids, in: NMR basic principles and progress, Vol. **3** (Springer, Berlin).

Part B

Summary of
Experimental Results

4 | COLOUR CENTRES IN THE ALKALI HALIDES

Colour centres in alkali halides belong to the best known defects in solids. Since the early work of Pohl and his coworkers in Göttingen, a huge amount of information has been achieved using a wide variety of methods. These include measurements of optical absorption, luminescence, ionic and photoconductivity and, in the last 20 years, spin resonance. The resonance methods have been able to prove beyond all doubt the formerly accepted model of the F centre. For other centres (M, R), decisions for one or the other of the proposed models could only be achieved by resonance methods, and some new centres (V_K) have been added to the colour centre family. The ESR and ENDOR results are often very instructive examples for the ability of these methods to analyse the structure of defects in solids (see section 2.2).

For an introduction to the general properties of colour centres the reader is referred to the book by Schulman and Compton (1962), to a review article by Pick (1965) and to a new book edited by Fowler (Fowler, 1968) which contains eight contributions by competent authors covering the latest developments of colour centre physics. A contribution of Seidel and Wolf (1968), 'ESR and ENDOR spectroscopy of colour centres in alkali halide crystals', is a modernized English version of a review article by the same authors (Seidel and Wolf, 1965), written in German in 1964. This article gives both a good introduction to the subject with thoroughly discussed examples and a complete covering of literature up to 1966. Therefore only some typical results of resonance work on colour centres shall be dealt with in this book, and the reader is referred to the Seidel-Wolf article for further study.

4.1. The F centre

F centres in alkali halides can be generated by irradiation (UV, X-ray, fast particles), by electrolysis and, most homogeneously and without accompanying production of other centres, by heating the crystals in alkali metal vapour. These crystals are coloured by an absorption band in the visible part of the spectrum. NaCl with F centres ('Farbzentren') looks yellow, KCl violet and KBr blue. According to De Boer, the F centre is an anion vacancy with an excess electron (fig. 4.1).

The F centre is paramagnetic, which has been confirmed by susceptibility measurements. In ESR experiments one expects an HFS interaction, especially with the nuclei of the surrounding ions. Since, e.g. in KCl, the six neighbouring K nuclei have spins $I = \frac{3}{2}$, one expects a spectrum with 19 HFS lines ($M = -9, \cdots, 0, \cdots, +9$) and intensity ratios of $1 : 6 : 21 : 56 : 120 : 216 : 336 : 456 : 564 : 580 : 546 : \cdots$. Fig. 4.2 shows the ESR spectrum in KCl; no HFS is resolved, the line width is 47 G.

This has been explained as follows: The observed line is the envelope of a huge number of unresolved HFS components, to which contribute the interactions with the nuclei not only in the nearest shell but also in the second and following shells. In fig. 4.3, the splitting due to the interaction with the 6 nuclei of shell I is shown in the upper part. Each of these lines is further split by interaction with the 12 nuclei of shell II, which is sketched in the lower part of fig. 4.3 only for the component $M_I = -2$.

Similar inhomogeneously broadened ESR lines have been observed in LiCl, NaCl, NaBr, KF, KBr, KJ and RbBr. In other crystals (LiF, NaF, RbCl, CsCl), the HFS interactions can partially be resolved. The expected 19-line spectrum with the contributions of the first shell neighbours has been observed in NaH. Some data are given in table 4.1.

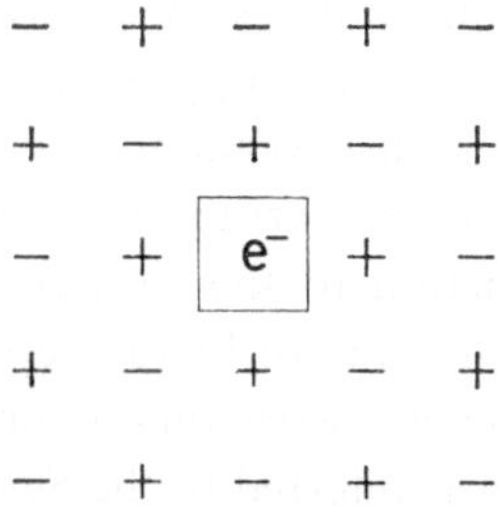

Fig. 4.1. F centre in a NaCl type lattice.

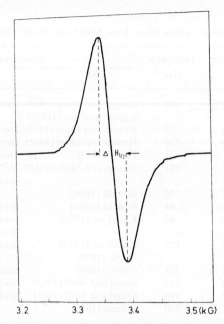

Fig. 4.2. ESR of F centres in KCl. Recorded is the first derivative of the absorption ($\nu = 9.4\,\text{GHz}$, $T = 300\,\text{K}$). (After Pick, 1965.)

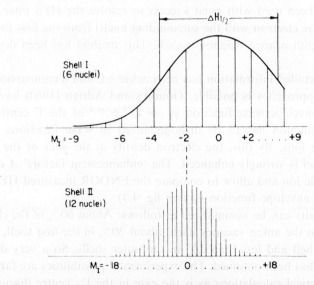

Fig. 4.3. The unresolved hyperfine structure of the F centre resonance in KCl. (After Seidel and Wolf, 1968.)

TABLE 4.1

F centres in alkali halides. (Data taken from Seidel and Wolf, 1968.)

Crystal	g factor	Halfwidth (G)	References	
			ESR	ENDOR
LiF	2.002	150	Holton and Blum (1962) Kaplan and Bray (1963)	Holton and Blum (1962)
LiCl	2.002	53	Holton and Blum (1962)	Holton and Blum (1962)
NaF	2.000	220	Holton and Blum (1962) Doyle (1963)	Holton and Blum (1962) Doyle (1963)
NaCl	1.998	145	Holton and Blum (1962)	Holton and Blum (1962) Seidel (1961)
NaBr	1.985	250	Schmid (1966)	–
KF	1.996	91	Schmid (1966)	Seidel (1961)
KCl	1.996	47	Kip et al. (1953)	Seidel (1961) Kersten (1968)
KBr	1.983	125	Kip et al. (1953) Noble (1959)	Seidel (1961)
KJ	1.965	225	Noble (1959)	Seidel (1961)
RbCl	1.980	425	Seidel and Wolf (1963)	Seidel and Wolf (1963)
RbBr	1.967	390	Seidel and Wolf (1963)	Seidel and Wolf (1963)
CsCl	1.97	700	Schmid (1966)	

While the unresolved ESR spectra of most F centres give only poor information concerning the structure of these centres, the powerful ENDOR method has been used with great success to resolve the HFS interactions of the F centre electron with the surrounding nuclei from the first to sometimes the eighth nearest neighbour shell. This method has been described in chapter 2.

Much detailed information has been achieved, and a comparison with theoretical approaches is possible. Gourary and Adrian (1960) have constructed an envelope wave function $\psi_F \approx r^{-1} e^{-\pi r/d}$ of the F centre electron. This function has to be orthogonalized to wave functions of the neighbouring ions. By this, the electron density at the sites of the neighbouring nuclei is strongly enhanced. The 'enhancement factors' A depend on the specific ion and allow to compare the ENDOR measured HFS constants to the envelope function ψ_F (see fig. 4.4).

The results can be summarized as follows: About 60% of the electron is localized in the anion vacancy itself, about 30% in the first shell, 6% in the second shell and less than 1% in the outer shells. So a very detailed information has been obtained. The experimental possibilities are far ahead of the theoretical calculations as is the case in the U_2 centre discussed in chapter 2.2.

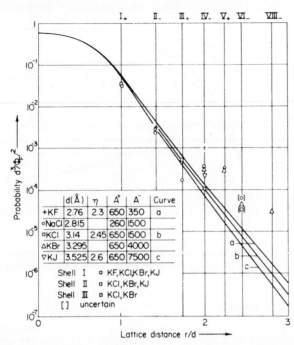

Fig. 4.4. Electron density in the neighbourhood of an F centre. (After Seidel, 1961.)

Some further results on F centres shall only be mentioned shortly. The F centre ESR has a pronounced saturation behaviour which has been subject to many investigations. The absorption signal saturates somewhat between the proposed behaviour of a homogeneously broadened line (Bloch) and the expected behaviour of an inhomogeneously broadened one (Portis). This is shown in fig. 4.5. The dispersion signal does not saturate, therefore this signal is easier to observe especially at low temperatures.

The relaxation times, in the form $(T_1 T_2)^{1/2}$, can be determined from saturation measurements. Determination of T_1 and T_2 separately has in some cases been done by 'burning a hole' into the broadened line. The most direct method for measuring T_1 and T_2, the spin–echo method, has so far not been applied here.

Recently, g factor and linewidth of the excited state of the F centre were determined by Mollenauer et al. (1969) using a special technique combining optical pumping, optical detection and resonating microwaves. These experiments represents a record for ESR sensitivity, the smallest measurable excited-state population being about 1000 spins for 1 G ESR linewidth.

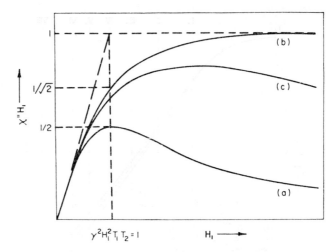

Fig. 4.5. ESR saturation behaviour of (a) a homogeneous line (Bloch, 1946), (b) an inhomogeneously broadened line (Portis), and (c) the F centre line. (After Seidel and Wolf, 1968.)

4.2. F aggregate centres

By irradiation with light or by annealing, F centres are converted into F aggregate centres ('F′ centre'). The F′ centre is a vacancy with two electrons and not paramagnetic in its ground state. The same holds for the M centre, which consists of two nearest neighbour F centres along one of the (110) directions. No ESR spectrum has been found in its ground state.

However, by irradiation an excited triplet state can be formed which decays with a time constant of 50 s at 90 K. Seidel (1963b) succeeded in obtaining ESR and ENDOR spectra from this excited state. They could be described by a spin Hamiltonian with $S = 1$ and the appropriate symmetry. Measurements of the excited state of the R centre, which consists of three nearest neighbour F centres in a (111) plane could be described with $S = \frac{3}{2}$ (quartet state). By these impressive measurements on the excited states, the models proposed earlier by Van Doorn and Haven (1956) and Pick (1960) could be confirmed.

Prolonged irradiation with neutrons results in higher aggregated centres and finally in the formation of alkali colloids. Fig. 4.6 shows this process with neutron bombarded LiF. The resonance of isolated F centres (fig. 4.6a) gets narrower with further irradiation, the signal being composed of the F line and aggregate centre lines, and finally the ESR of conduction electrons in Li colloids can be seen.

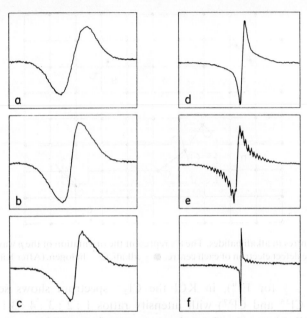

Fig. 4.6. Variation of ESR spectra in neutron irradiated LiF with neutron dose. The neutron dose varies from (a) 10^{16} cm^{-2} to (d,e,f) 5×10^{18} cm^{-2}. (After Kaplan and Bray, 1963.)

4.3. V centres

Centres with a defect electron (hole) are called V centres, because the known optical absorption bands lie on the violet side of the F bands. These centres are most commonly generated by irradiation with X-rays. The earlier opinion that these centres were the antimorphs to the electron excess centres (i.e. F and F aggregate centres) had to be abandoned by the spin resonance results. Especially, no 'anti F centre', which would be a hole in a cation vacancy, seems to exist.

These important results are mostly due to Känzig (1960) and coworkers. Fig. 4.7 summarizes their work and gives the models for (A) the V_K centre, (B) the H centre, (C) the V_F centre and (D) the V_t centre, which were obtained by carefully analysing the HFS interaction in the normal ESR spectra.

For example, the V_K centre has a hole shared by two neighbour anions which are displaced toward one another along a (110) direction. This X_2^- molecule ion shows a pronounced hyperfine spectrum due to two equivalent nuclei. In LiF, the F_2^- spectrum has three lines with intensity ratios

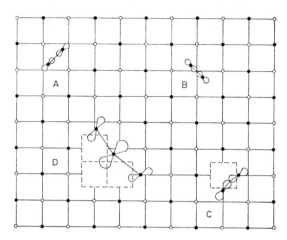

Fig. 4.7. V centres in alkali halides. The 8's represent the orientation of the p wave functions describing the defect electron of each centre. ● — alkali, ○ — halogen. (After Känzig, 1960.)

$1 : 2 : 1$ ($I = \frac{1}{2}$ for F^{19}), in KCl the Cl_2^- spectrum shows seven lines ($I = \frac{3}{2}$ for Cl^{35} and Cl^{37}) with intensity ratios $1 : 2 : 3 : 4 : 3 : 2 : 1$ (fig. 4.8). Complications arise from the presence of two Cl isotopes and from the fact that the magnetic field direction has different angles with the six possible directions of the centre axis (face diagonal) in the crystal (see section 1.6).

Important for the interpretation of X-ray colouring of alkali halides is the H centre. This centre can be regarded as an X_2^- molecule ion in place

H ‖ [100]

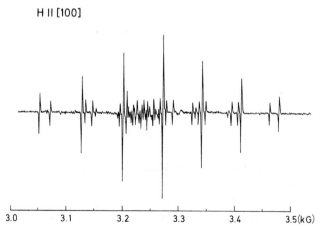

$$3.0 \qquad 3.1 \qquad 3.2 \qquad 3.3 \qquad 3.4 \qquad 3.5\,(kG)$$

Fig. 4.8. ESR of V_K centres in KCl ($\nu = 9.263$ GHz). (After Castner and Känzig, 1957.)

of an X^- anion. Chemically it is equivalent to an interstitial halide atom and therefore complementary to the F centre, which chemically is an excess alkali atom. Both centres are formed together during low temperature X-ray irradiation. Recombination of pairs of F and H centres restores the lattice. An H centre adjacent to a substitutional cation impurity was recently identified by Patten and Keller (1969) as being responsible for the well known V_1 optical absorption band.

The V_F centre could be the 'antimorph' of the F centre, consisting of a hole in a cation vacancy. But its symmetry and wave function are quite different: it is not cubic and can best be described as a V_K centre in the neighbourhood of a vacancy.

4.4. Impurity centres

All crystals contain impurities, either wanted or unwanted. The high sensitivity of the ESR technique can be used to analyse crystals for the presence of impurities. If these centres are not paramagnetic, often transformations into paramagnetic ones can be achieved by suitable processes.

The U_2 centre (hydrogen atom at an interstitial site) has been discussed in detail in section 2.2 because of its pronounced HFS and super HFS spectrum. At higher temperatures, the H atoms become mobile and sometimes react with other, perhaps diamagnetic, impurities to form paramagnetic ones. An example is given in fig. 4.9 which shows the resonance of an HCN^- centre formed in this way in KCl crystals containing CN^- impurities.

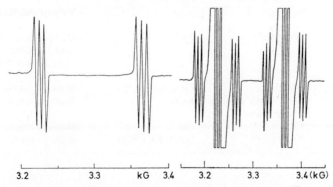

Fig. 4.9. ESR of HCN^- centres in KCl. With extreme sensitivity, the C^{13} hyperfine structure is additionally recorded in the right hand figure. (After Beuermann and Hausmann, 1967.)

TABLE 4.2

Impurity centres in alkali halides

Impurity	Centre	Crystal	References
Hydrogen	$H_i{}^0$ (U_2)	KCl, KBr, KJ, NaCl	Delbecq et al. (1956) Kerkhoff et al. (1963) Spaeth (1966, 1968, 1969) Hausmann and Köpp (1971)
	U_3, U_4	KCl, NaCl	Hayes and Hodby (1966) Bessent et al. (1965)
	$H_i{}^0$	CaF_2	Hall and Schumacher (1962)
Nitrogen	NO, $NO_2{}^{2-}$, $NO_3{}^{2-}$	KCl, KBr, KJ	Jaccard (1961) Schoemaker and Boesman (1963)
	$N_2{}^{2-}$	KCl	Schoemaker and Boesman (1963)
	$N_2{}^-$	KCl, KBr, KJ, NaCl	Sander (1962) Hausmann et al. (1964) Brailsford et al. (1968a)
	N^{2-}	KCl	Hausmann (1966a)
Oxygen	O^-	KCl, KBr, KJ, RbCl, RbBr, NaJ	Sander (1962, 1964) Brailsford et al. (1968b, 1969)
	$O_2{}^-$	KCl, KBr, KJ, RbCl, RbBr, RbJ	Känzig and Cohen (1959) Zeller and Känzig (1967) Doyle and Castner (1968)
	$O_2{}^+$	KCl, KBr	Jaccard (1961)
	$S_3{}^-$, $Se_3{}^-$	KCl, NaCl	Schneider et al. (1966) Suwalski and Seidel (1966)
Sulphur, selenium	$SO_2{}^-$, $SeO_2{}^-$ S^-, Se^-, H_2S^- $S_2{}^-$	KCl, KBr	Hausmann (1966b) Vanotti and Morton (1967)
Silver	Ag^0, Ag^{2+}	KCl	Delbecq et al. (1963) Seidel (1963a) Mallick and Schumacher (1966)
Manganese	Mn^{2+}	LiCl, NaCl	Watkins (1959)
	Mn^{2+} with vacancy aggregates	KCl	Watkins (1959)
Chromium	Cr^+	NaCl	Welber (1965)
Europium	Eu^{2+}	NaCl, KCl, KBr	Röhrig (1965)

Oxygen ($O_2{}^-$, O^-), nitrogen (NO, $NO_2{}^{2-}$, $NO_3{}^{2-}$, $N_2{}^{2-}$, $N_2{}^-$) and sulphur centres should be mentioned here because most crystals grown under normal atmospheric conditions contain such impurities and ESR measurements can be used to analyse the crystal purity.

Our knowledge about the behaviour of divalent cations (Ca^{2+}, Ba^{2+}, Mg^{2+}) stems mostly from the analysis of ionic conductivity and diffusion experiments. Because of electrical neutrality, each crystal must contain as many cation vacancies as divalent cation impurities ions. These vacancies are isolated from the divalent ions at high temperatures and associated with them at lower temperatures. Watkins (1959) was able to observe this association by using Mn^{2+} impurities in LiCl crystals. The ESR spectra are different for isolated Mn^{2+} ions, Mn^{2+} vacancy associates and Mn aggregates. Results on impurity centres are summarized in table 4.2.

References

Bessent, R. G., W. Hayes and J. W. Hodby, 1965, Phys. Letters **15**, 115.
Beuermann, G. and A. Hausmann, 1967, Z. Physik **204**, 425.
Bloch, F., 1946, Phys. Rev. **70**, 460.
Brailsford, J. R. and J. R. Morton, 1969, J. Chem. Phys. **51**, 4794.
Brailsford, J. R., J. R. Morton and L. E. Vanotti, 1968a, in: Proc. Color Center Symp. Rome, 28.
Brailsford, J. R., J. R. Morton and L. E. Vanotti, 1968b, J. Chem. Phys. **49**, 2237.
Castner, T. G. and W. Känzig, 1957, J. Phys. Chem. Solids **3**, 178.
Delbecq, C. J., B. Smaller and P. H. Yuster, 1956, Phys. Rev. **104**, 599.
Delbecq, C. J., W. Hayes, M. C. M. O'Brien and P. H. Yuster, 1963, Proc. Roy. Soc. London A **271**, 243.
Doyle, A. M. and T. G. Castner, 1968, in: Proc. Color Center Symp. Rome, 54.
Doyle, W. T., 1963, Phys. Rev. **131**, 555.
Fowler, W. B., ed., 1968, Physics of color centres (Academic Press, New York).
Gourary, B. S. and F. J. Adrian, 1960, Solid State Phys. **10**, 127.
Hall, J. L. and R. T. Schumacher, 1962, Phys. Rev. **127**, 1892.
Hausmann, A., 1966a, Kurznachr. Akad. Wiss. Göttingen **20**, 91.
Hausmann, A., 1966b, Z. Physik **192**, 313.
Hausmann, A. and S. Köpp, 1971, Z. Physik **243**, 382.
Hausmann, A., R. Hilsch and W. Sander, 1964, Z. Physik **179**, 461.
Hayes, W. and J. W. Hodby, 1966, Proc. Roy. Soc. London A **294**, 359.
Holten, W. C. and H. Blum, 1962, Phys. Rev. **125**, 89.
Jaccard, C., 1961, Phys. Rev. **124**, 60.
Känzig, W., 1960, J. Phys. Chem. Solids **17**, 88.
Känzig, W. and M. H. Cohen, 1959, Phys. Rev. Letters **3**, 509.
Kaplan, R. and P. J. Bray, 1963, Phys. Rev. **129**, 1919.
Kerkhoff, F., W. Martienssen and W. Sander, 1963, Z. Physik **173**, 184.
Kersten, R. 1968, Phys. Stat. Sol. **29**, 575.
Kip, A. F., C. Kittel, R. A. Levy and A. M. Portis, 1953, Phys. Rev. **91**, 1066.

Mallick, G. T. and R. T. Schumacher, 1966, Bull. Am. Phys. Soc. 11, 171.
Mollenauer, L. F., S. Pan and S. Yngresson, 1969, Phys. Rev. Letters 23, 683.
Noble, G. A., 1959, J. Chem. Phys. 31, 931.
Patten, F. W. and F. J. Keller, 1969, Phys. Rev. 187, 1120.
Pick, H., 1960, Z. Physik 159, 69.
Pick, H., 1965, Springer tracts in modern physics, Vol. 38 (Springer, Heidelberg).
Röhrig, R., 1965, Phys. Letters 16, 20.
Sander, W., 1962, Z. Physik 169, 353.
Sander, W., 1964, Naturwissenschaften 51, 404.
Schmid, D., 1966, Phys. Stat. Sol. 18, 653.
Schneider, J., B. Dischler and A. Räuber, 1966, Phys. Stat. Sol. 13, 141.
Schoemaker, D. and E. Boesman, 1963, Phys. Stat. Sol. 3, 1695.
Schulman, J. H. and W. D. Compton, 1962, Color centers in solids (Pergamon, Oxford).
Seidel, H., 1961, Z. Physik 165, 218, 239.
Seidel, H., 1963, Phys. Letters 6, 150.
Seidel, H., 1963, Phys. Letters 7, 27.
Seidel, H. and H. C. Wolf, 1963, Z. Physik 173, 455.
Seidel, H. and H. C. Wolf, 1965, Phys. Stat. Sol. 11, 3.
Seidel, H. and H. C. Wolf, 1968, in: W. B. Fowler, ed., Physics of color centers (Academic Press, New York).
Spaeth, J. M., 1966, Z. Physik 192, 107.
Spaeth, J. M., 1968, in: Proc. Color Center Symp. Rome, 177.
Spaeth, J. M., 1969, Phys. Stat. Sol 34, 171.
Suwalski, J. and H. Seidel, 1966, Phys. Stat. Sol. 13, 159.
Van Doorn, C. Z. and Y. Haven, 1956, Philips Res. Rept. 11, 419.
Vannotti, L. E. and J. R. Morton, 1967, Phys. Rev. 161, 282.
Watkins, G. W., 1959, Phys. Rev. 113, 79, 91.
Welber, B., 1965, Phys. Rev. 138, A 1481.
Zeller, H. R. and W. Känzig, 1967, Helv. Phys. Acta 40, 845.

5 PARAMAGNETIC DEFECTS IN II–VI GROUP COMPOUNDS

Zinc sulphide has been the first semiconductor to be studied by spin resonance as early as 1951; however, most measurements on II–VI compounds have been performed during the last years. The main reason for this is the fact that the preparation of single crystals with controlled impurity content has remained a major problem. The investigations that have been done up to 1961 were reviewed in an article by Ludwig and Woodbury (1962) concerning ESR measurements in semiconductors in general. More data obtained till 1965 have been summarized in a review by Title (1967).

The II–IV compounds of interest are a somewhat intermediate case between insulators and elemental semiconductors such as silicon and germanium. The bonding is a mixture of covalent and ionic bonding, and the energies of the band gaps are between those of pure insulators and those of elemental semiconductors. The compounds show a tetrahedral arrangement of atoms: Each metal atom is at the centre of a tetrahedron of group VI atoms and vice versa. Many substances crystallize in a cubic lattice (ZnSe, ZnTe and CdTe), while some have a hexagonal structure (ZnO, CdS); others have both a cubic form (sphalerite) and a hexagonal form (wurtzite), e.g. ZnS.

No matter which compound we consider, we have to deal with a diamagnetic substance as the spins of the bonding electrons are paired so that the total net spin of the electrons is zero. Therefore II–VI compounds are very suitable substances for investigating the environment of a paramagnetic impurity or defect in a solid.

The defects in question are of various kinds. Most of them are either substitutional or interstitial impurities including transition metal ions of the

3d group or rare earth ions of the 4f group, the substitution being either at the cation site or at the anion site and often involving cation or anion vacancies. Other paramagnetic defects are shallow donor and acceptor centres, or missing bonding electrons or holes.

In all the II–VI compounds one finds a great tendency towards compensation of the unpaired bonding electrons in a defect, thus making it difficult or even impossible to observe paramagnetism. For this case, a useful technique has been introduced by Lambe et al. (1959) and Title (1959) by using optical excitation to add or to remove an electron at a defect. This can be done either by light of energy greater than the band gap creating electron–hole pairs or by direct excitation of an electron from the valence band to a defect by light of energy less than the band gap. In the latter case, the remaining hole may move in the valence band and be trapped by another defect. A great number of optical processes have meanwhile been understood by using resonance spectra as a monitor for defect centres which become paramagnetic by optical irradiation.

5.1. The Hamiltonian

The spectra of substitutional impurities generally display the full symmetry of the substitutional site. In cubic crystals, the spin Hamiltonian has the following form

$$H = g\,\mu_B\,\boldsymbol{H}\cdot\boldsymbol{S} + A\,\boldsymbol{S}\cdot\boldsymbol{I} + \tfrac{1}{6}\,a\,(S_x^4 + S_y^4 + S_z^4) + A_j\,\boldsymbol{S}\cdot\boldsymbol{I}_j$$
$$+ P_j\,\boldsymbol{I}_j\cdot\boldsymbol{I}_j - g_{Ij}\,\mu_N\,\boldsymbol{H}\cdot\boldsymbol{I}_j. \tag{5.1}$$

Here A_j, P_j, g_{Ij} and I_j refer to the nucleus of the jth neighbouring atom with which there is nonzero interaction.

For centres in an axial crystal field, i.e. in hexagonal lattices, the spin Hamiltonian has the form of eq. (1.13), augmented with hyperfine terms containing interactions with neighbouring nuclei similar to those in eq. (5.1). The polar axis c is the axis of symmetry.

For impurities with $S = \tfrac{3}{2}$ in a cubic field, an anisotropic resonance is observed which can be explained by the addition of terms containing the cube of the spin operator multiplied by the field (Bleaney, 1959):

$$H = g\,\mu_B\,\boldsymbol{S}\cdot\boldsymbol{H} + u\,[(S_x^3\,H_x + S_y^3\,H_y + S_z^3\,H_z)$$
$$- \tfrac{1}{5}\,\boldsymbol{S}\cdot\boldsymbol{H}\,\{3S\,(S + 1) - 1\}]$$
$$+ A\,\boldsymbol{I}\cdot\boldsymbol{S} + U\,[(S_x^3\,I_x + S_y^3\,I_y + S_z^3\,I_z)$$
$$- \tfrac{1}{5}\,\boldsymbol{I}\cdot\boldsymbol{S}\,\{3S\,(S + 1) - 1\}]. \tag{5.2}$$

Similar terms in U are used to explain the anisotropic hyperfine structure.

An additional complication in the hexagonal crystals is the presence of two geometrially nonequivalent sites whose Hamiltonians differ as to the orientation of the x', y', z' axes of eq. (1.13). When a cubic field splitting term is present, the two sites are magnetically distinguishable, giving rise to a splitting of all lines proportional to $a \sin^3 \vartheta \cos \vartheta \cos 3\psi$. Here ψ is the angle between the magnetic field and the c axis of the crystal and ϑ is the angle of rotation around the c axis. This splitting can be used to determine the cubic field parameter directly (Hausmann, 1968). For the orientation $H \parallel$ c and $H \perp$ c, the line positions are only sensitive to the combination $|a - F|$. In fig. 5.1, the splitting of the transitions versus the angle ϑ together with $a |\sin^3 \vartheta \cos \vartheta \cos 3\psi|$ is plotted considering higher order corrections in a.

Because of the two magnetically distinguishable lattice sites, the macroscopic crystal growth behaviour of the hexagonal crystals can be determined. This can be seen as follows: The outer surfaces of the crystals can be built up in two possible ways. In fig. 5.2, the projections of two neighbouring planes normal to c are shown, each of which can be obtained from the other one by a rotation of 30° about c. Maximum splitting of the lines is observed when the directions of c, the field H and the $\langle 100 \rangle$ axis of one of the two possible lattice sites are in one plane. The experiments show that maximum splitting occurs if the field H is normal to the surfaces of the crystals. However, only in fig. 5.2a the angle ψ between the planes containing c,$\langle 100 \rangle$ and H,$\langle 100 \rangle$, respectively, equals zero. So it can be concluded that the

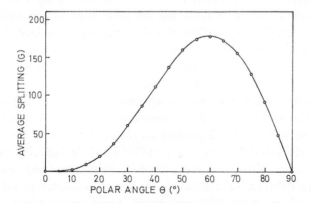

Fig. 5.1. Polar variation of the average splitting of the transition $M = \pm \, ^5/_2 \leftrightarrow \pm \, ^3/_2$ for ZnO : Fe^{3+}. The solid curve represents the function $(25\sqrt{2}/_3)\, a\, |\cos \theta \sin^3 \theta \cos \psi|$ for $\psi = 0°$ and $|a| = 46.8$ G.

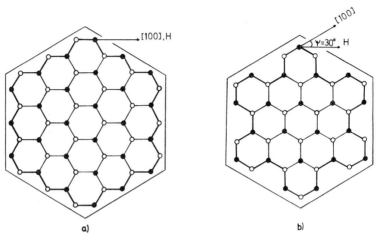

Fig. 5.2. Sketch of the two possible ways of crystal growth in a plane normal to c.
● – Zn^{2+} ion, ○ – O^{2-} ion.

crystals grow as indicated in fig. 5.2(a) rather than as in fig. 5.2(b). Thus the surfaces are formed by closer packed planes as is expected from arguments of minimum energy.

5.2. Iron group impurities

ESR parameters of individual 3d ions are represented in tables 5.1 and 5.2. The impurities are labelled by their electron configuration rather than by their charge state. This simplifies the comparison of the ESR results in the II–VI compounds with those for the same impurities in different ionic compounds. The electron configuration does not imply a defined charge state in the compound as this depends on whether one thinks in a covalent or an ionic model. The numerous results on 3d ions have been discussed in detail by Title (1967) and in the original papers.

As far as known, the iron group ions substitute for the column II elements. Many of the impurity centres are sensitive to illumination with ultraviolet or visible light, depending on the charge state of the ion incorporated. Numerous defects can act as electron or hole traps after irradiation. The investigation of the same ion in different II–VI compounds may be used to study the influence of ionic radii, change of covalency or other lattice dependent effects on the ESR parameters. Some special topics concerned with the incorporation of iron group impurities in II–VI semiconductors are discussed below.

TABLE 5.1

ESR parameters for iron group impurities in the cubic II–VI compounds

Ion	Crystal	g	a $(10^{-4}$ $cm^{-1})$	A $(10^{-4}$ $cm^{-1})$	T (K)	Photo-sensi-tive	References
V^{3+} (d^2)	ZnS	1.9433	–	63.0	1.3	yes	Holton et al. (1964)
	ZnTe	1.917	–	57.8	–	–	Woodbury and Ludwig (1961)
Ti^{2+}	ZnS	1.928	–	12.8	77	yes	Schneider and Räuber (1966)
Cr^{3+} (d^3)	ZnSe	2.317	–	–	77	–	Rai et al. (1967)
	ZnTe	3.28	–	–	10	–	Woodbury and Ludwig (1961)
Cr^{2+} (d^4)	ZnSe	$g_{\parallel}=7.787$ $g_{\perp}=0$	–	12.6	1.3	–	De Wit et al. (1965)
Cr^+ (d^5)	ZnS	1.9995	3.9	13.4	77	yes	Dieleman (1962)
	ZnSe	2.0016	5.35	13.3	77	yes	Title (1964b)
	ZnTe	2.0023	6.60	12.4	77	yes	Title (1964b)
	CdTe	1.9997	3.1	12.8	4.2	–	Ludwig and Lorenz (1963)
Mn^{2+}	ZnS	2.0022	7.94	−64.5	77	–	Schneider et al. (1963b)
	ZnSe	2.0051	19.7	−61.7	10	–	Schneider et al. (1963b)
	ZnTe	2.0105	30	−56.2	20	–	Woodbury and Ludwig (1961)
	CdTe	2.0075	27	−57.1	77	–	Woodbury and Ludwig (1961)
Fe^{3+}	ZnS	2.0194	127.4	7.69	77	yes	Räuber and Schneider (1962a)
	ZnSe	2.0469	48.3	6.75	77	yes	Estle and Holton (1965)
	ZnTe	2.0969	2613	4.2	1.2	–	Henzel (1964)
Fe^+ (d^5)	ZnS	2.2515	–	–	1.3	yes	Holton et al. (1964)
Co^{2+}	ZnS	2.2742	–	7.7	4.2	yes	Hennig et al. (1966)
Ni^{3+}	ZnS	2.1480	–	–	1.3	yes	Holton et al. (1964)
Ni^{3+}	ZnSe	2.1978	–	–	1.3	yes	Watts (1969)
Ni^+	ZnSe	1.4374	–	81.4	1.3	yes	Watts (1969)

5.2.1. SUPERHYPERFINE STRUCTURE

One of the characteristics of iron group transition ions resulting from their incorporation as impurities into a crystalline environment is the resultant delocalization of the electronic orbitals. This delocalization leads to hyperfine interaction with nuclei neighbouring the impurity which is much larger than calculated for point dipole interactions. These interactions with nearby nuclei manifest themselves in an additional splitting of the resonance spectrum, often referred to as superhyperfine structure (SHFS). (A first example for SHFS, the U_2 centre, has been treated in chapter 2.) The SHFS is a measure for the spread of the wave function of the impurity ion. Early measurements showed that the small covalent bonding present even in ionic crystals dominates the SHFS (Owen and Stevens, 1953). Therefore it is of interest to explore the characteristics of iron group impurities in more covalent crystals, such as the II–VI compounds.

A number of ESR observations of SHFS associated with 3d impurities were reported in the last few years. Cadmium SHFS has been observed for

TABLE 5.2

ESR parameters for iron group impurities in the hexagonal II–VI compounds

Ion	Crystal	g	$g_{\parallel}$	$g_{\perp}$	D (10^{-4} cm^{-1})	$\lvert a-F\rvert$ (10^{-4} cm^{-1})	a (10^{-4} cm^{-1})	$A_{\parallel}$ (10^{-4} cm^{-1})	$A_{\perp}$ (10^{-4} cm^{-1})	Photo-sensitive	T (K)	References
Ti^{2+} (d^2)	ZnS	—	1.929	1.926	125	—	—	—	—	—	—	Schneider (unpublished)
V^{3+}	CdS	—	1.923	1.921	480	—	—	13.5	—	—	77	Dziesiaty and Böttcher (1968)
	CdS	—	1.934	1.932	1130	—	—	63	66	—	10	Woodbury and Ludwig (1961)
Cr^{2+} (d^4)	CdS	—	7.747	0	1503	—	—	—	—	—	—	Morigaki (1963a)
	ZnSe	—	7.837	0	—	—	—	13.8	—	yes	1.3	Estle and Holton (1966)
Mn^{2+} (d^5)	ZnO	—	1.9984	1.9998	252.2	5.83	7.1	79.2	78.7	—	300	Hausmann and Huppertz (1968)
	ZnS	—	2.0018	2.0018	130.9	7.7	7.35	64.9	64.9	—	300	Schneider et al. (1963b)
	ZnSe	—	2.0055	2.0055	425.1	11.85	17.66	61.2	61.2	—	300	Estle and Holton (1966)
	ZnTe	2.0057	—	—	—	—	—	61.8	—	—	1.3	Estle and Holton (1966)
	CdS	2.0020	—	—	8.2	4.2	3.3	66.0	—	—	77	Lambe and Kikuchi (1960)
	CdSe	2.0041	—	—	15.6	16.5	14.3	62.2	—	—	77	Schneider et al. (1963b)
	CdTe	2.0069	—	—	—	—	27.7	57.5	—	—	1.5	Estle and Holton (1966)
Fe^{3+}	ZnO	—	2.0056	2.0041	587	37	46.8	—	—	yes	300	Hausmann (1969a)
	ZnSe	2.0470	—	—	—	—	45	—	—	—	1.3	Estle and Holton (1966)
	ZnTe	—	—	—	—	—	—	—	—	—	1.3	Estle and Holton (1966)
	CdS	2.01	—	—	30	30	30	—	—	—	4.2	Morigaki and Hoshina (1965)
Co^{2+} (d^7)	ZnO	—	2.2500	4.5536	—	—	—	16.1	6.0	—	20	Hausmann (1969b)
	CdS	—	2.269	2.279	1300	—	—	4.7	9.1	yes	1.5	Morigaki (1963b)
	CdSe	—	2.295	2.303	770	—	—	13.6	20.1	yes	1.5	Hoshina (1966)
Ni^{3+}	ZnO	—	2.1426	4.1379	—	—	—	—	—	yes	1.5	Holton et al. (1964)
Ni^{2+} (d^8)	CdS	—	2.13	5.1	—	—	—	—	—	yes	1.5	Morigaki (1964b)
Cu^{2+} (d^9)	BeO	—	1.709	2.379	—	—	—	50	108	—	1.4	De Wit and Reinberg (1967)
	ZnO	—	0.7392	1.5182	—	—	—	$A^{63}=198$, $A^{65}=210$	$B^{63}=228$, $B^{65}=245$	—	1.3	Hausmann and Schreiber (1969)
	CdS	—	2.240	1.75	—	—	—	99	—	yes	1.5	Morigaki (1964a)

Mn^{2+} in CdSe, CdS and CdTe; Cr^+ in CdTe; Co^{2+}, Ti^{2+}, Cr^{2+} and V^{3+} in CdS. Zinc SHFS was observed for Cr^+ and Mn^{2+} in ZnS. Crystals of ZnSe and ZnTe provide a good opportunity to observe Se or Te SHFS because of the low relative abundance of Zn isotopes with a nuclear moment and the moderate abundance of Se and Te isotopes with a nuclear moment. SHFS with one or more shells of the chalcogenides has been observed for Cr^+, Mn^{2+}, Fe^{3+}, Cr^{2+} and Fe^+ in ZnSe and ZnTe host lattices. Data on SHFS are tabulated in table 5.3. The index of the parameters indicates the number of the neighbouring shell. Fig. 5.3 shows for example the SHFS of V^{51} in CdS, associated with 13% abundant Cd^{111} and 12% abundant Cd^{113}, which have nuclear spin $\frac{1}{2}$ and approximately equal nuclear moments. SHFS with the 12 equivalent nearest Cd sites is resolved (Woodbury and Ludwig, 1961).

If SHFS is observed, the spin Hamiltonian in eq. (1.13) has to be augmented with the term

$$+ \sum_n [A_n \, S \cdot I_n + g_I \, \mu_N \, H \cdot I_n],$$

where the sum over n is taken over all nuclei having an appreciable interaction with the impurity spin.

TABLE 5.3
Superhyperfine structure parameters of iron group impurities in II–VI compounds (in units of 10^{-4} cm^{-1})

Ion	Crystal	$|A_1|$	$|A_2|$	$|A_1|$	$|B_1|$	a_1	b_1	References
Cr^+	ZnS(cubic)	≤ 2.2	≤ 1.3	–	≤ 2.2	–	–	Dieleman et al. (1962)
	ZnSe	6.83	≤ 0.5	1.65	4.41	–	–	Estle and Holton (1966)
	ZnTe	11.8	–	3.54	21.83	–	–	Estle and Holton (1966)
	CdTe	≤ 10	5.82	–	–	–	–	Ludwig and Lorenz (1963)
Mn^{2+}	ZnS	–	–	–	–	–	–	Schneider et al. (1963b)
	ZnSe	2.10	–	0.44	2.81	2.57	−0.24	Estle and Holton (1966)
	ZnTe	3.21	–	1.14	4.85	4.29	−0.54	Estle and Holton (1966)
	CdS	–	2.6	–	–	–	–	Lambe and Kikuchi (1960)
	CdSe	–	2.7	–	–	–	–	Schneider et al. (1963b)
	CdTe	–	2.6	–	–	–	–	Lambe and Kikuchi (1960)
Fe^{3+}	ZnSe	11.37	–	≤ 2.4	7.14	8.55	1.41	Estle and Holton (1966)
	ZnTe	22.0	–	3	16	18	2.0	Estle and Holton (1966)
	CdS	–	–	≤ 1.5	–	–	–	Estle and Holton (1966)
Fe^+	ZnTe	3.90	–	–	–	–	–	Estle and Holton (1966)
Co^{2+}	ZnSe	16.4	–	–	9.6	–	–	Hennig et al. (1966)
	CdS	–	1.92	–	–	–	–	De Kinder (1965)
Ti^{2+}	ZnS	–	2.0	–	–	–	–	Schneider and Räuber (1966)
V^{3+}	CdS	–	–	–	–	–	–	Woodbury and Ludwig (1961)

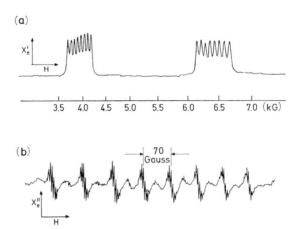

Fig. 5.3. The spectrum of V^{51} ($3d^2$) in CdS. (a). $\Delta M = 1$ transitions for H parallel to the c axis. (b). The $\Delta M = 2$ transition for the field $10°$ from the c axis. The superhyperfine structure of this transition is associated with 13% abundant Cd^{111} and 12% abundant Cd^{113} (which have nuclear spin $\frac{1}{2}$ and approximately equal nuclear moments) occupying the 12 equivalent nearest Cd sites. (After Ludwig and Woodbury, 1962.)

The amount of splitting resulting from SHFS interaction may be used for the calculation of the magnetic field at the site of the nucleus involved. By comparing ESR results of the same ion in different host lattices, qualitative conclusions concerning the spread of the wave function as well as information about distortions of the lattice in the environment of the impurity ion may be obtained. Such distortions are the reason for different line widths of the same centre in crystals of the same sort but of different quality. If the lattice is distorted, the crystal field is not identical for all ions. Thus the resonance field is slightly different, which generally leads to a broadening of the lines.

5.2.2. EFFECT OF COVALENCY ON THE ESR PARAMETERS

It has been shown for Mn^{2+} in a different host lattice that there is a deviation of the parameters g, D, a and A from crystal to crystal. Van Wieringen (1955) has shown that there is a regular variation with covalency.

The bonding in the II–VI compounds is a mixture of ionic and covalent bonding. Any increase in the covalency is accompanied by a corresponding decrease in ionicity. In covalent bonding, some of the electrons are transferred to the 4s Mn shell. This may diminuish the contribution of the configurational interaction within the $3s^2 3d^n$ configuration. In figs. 5.4a–c, the changes in g, D and a are plotted versus ionicity for Mn^{2+}. The variation

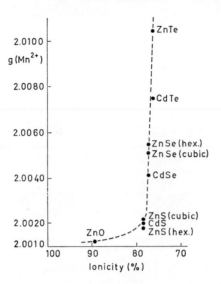

Fig. 5.4 (a). A plot of the g value of Mn^{2+} against the ionicity in the bonding. (After Title, 1967.)

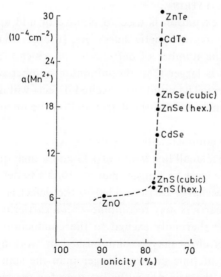

Fig. 5.4 (b). A plot of the a value of Mn^{2+} against the ionicity in the bonding. (After Title, 1967.)

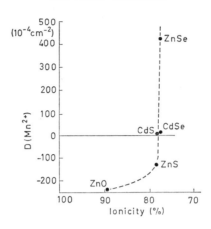

Fig. 5.4 (c). A plot of the D value of Mn^{2+} against the ionicity in the bonding. (After Title, 1967.)

of the parameters as a function of the change in ionicity has been calculated by Watanabe (1964), Van Wieringen (1955) and Matamura (1959). Good agreement with experiment is obtained.

5.2.3. OPTICAL EXCITATION

Often optical excitation is used to remove or add an electron or hole at an otherwise nonparamagnetic defect site. In order to use this method, it is necessary that the number of defect sites at which electrons have been added or removed is larger than the minimum number of spins detectable by the spectrometers. The number of excited defects will depend on the life-time of the electron or hole at the defect and on the intensity of the optical excitation.

For a normal commercially available X band spectrometer, such as in fig. 1.10, assuming a small line width of 1 G and a high quantum efficiency, the lifetime must at least be longer than 5×10^{-6} s to detect excited centres. The lifetime of the charge carrier added to the defect is influenced by two mechanisms. The carriers may recombine with a carrier of opposite sign, or the carrier may be thermally excited to the conduction or valence band. However, the latter effect may be diminuished by working at low tempera-tures. Further the lifetime must be longer than the spin lattice relaxation time, as otherwise there is not sufficient time for the spin to relax to its magnetic ground state and observation of resonance is impossible.

For a great number of 3d impurities it was found that ESR is sensitive to optical illumination. From measurements on photoluminescence it is known that Fe in ZnS, ZnSe or CdS may act as a 'killer' and thus reduce the efficiency of luminescence. Räuber and Schneider (1962a), for instance, found that the resonances of Fe^{3+} in ZnS are photosensitive and are connected with optical absorption and luminescence of self activated ZnS. It was concluded that hole transfer via the valence band could occur between Fe^{2+} ions and an association of a zinc vacancy and a donor. Thus Fe^{3+} ions are created and become recombination centres for electrons. In the case that the luminescence data can be correlated to ESR spectra, quantitative results can be obtained about where the centre is situated in the energy gap of the crystal. Detailed information on deep centre luminescence in the II–VI compounds can be found in chapter 9 of Aven and Prener (1967).

5.2.4. The Jahn–Teller effect of 3d ions

The transition metal ions played a crucial role in the development of the theory of the Jahn–Teller effect (see section 2.8). The 3d impurity point defects in insulating crystals are most conveniently studied when one is interested in this effect. Jahn–Teller distortions for nearly all charge states of the various 3d ions in different crystal field symmetries have been studied. As there is the excellent review by Sturge (1967), the reader is referred to that book for details and literature.

5.3. Rare earth impurities

During the last few years, quite a number of ESR measurements on rare earth impurities in II–VI compounds have been reported. Still, much less information is obtainable than for 3d ions. Therefore only limited conclusions can as yet be drawn from the measurements concerning the vicinity of the impurity.

The ESR of rare earth impurities in the zinc and cadmium chalcogenides and the parameters obtained by it are summarized in tables 5.4 and 5.5. In the following, the results are discussed for each of the rare earth ions so far observed.

Nd^{3+} ($4f^3$). Resonance of the Nd^{3+} ion in a trigonal field has been observed in CdS and CdSe. A spectrum due to isolated substitutional Nd^{3+} has been found in CdTe, where a trigonal Nd^{3+} resonance is observed if the CdTe is doped with Nd and P.

TABLE 5.4

ESR parameters of rare earth impurities in cubic II–VI compounds

Ion	Crystal	g	$\lvert b_4 \rvert$ [a]) $(10^{-4}$ $cm^{-1})$	$\lvert b_6 \rvert$ [a]) $(10^{-4}$ $cm^{-1})$	$\lvert A \rvert$ $(10^{-4}\ cm^{-1})$	T (K)	References
Eu^{2+} (f^7)	CdTe	1.9917	7.66	0.12	$A^{151} = 23.19$ $A^{153} = 10.25$	77	Title (1964a)
Gd^{3+} (f^7)	CdTe	1.9889	24.07	0.26	$A^{155} = 860.4$ $A^{155} = 238.0$	77	Title (1965)
Er^{3+} (f^{11})	ZnS	5.926	–	–	$A^{167} = 208$	1.3	Watts (1966)
	ZnSe	5.925	–	–	$A^{167} = 210$	1.3	Watts (1966)
	ZnTe	5.931	–	–	$A^{167} = 210$	1.3	Watts (1966)
	CdTe	5.941	–	–	$A^{167} = 210$	1.3	Watts (1966)
Ho^{2+} (f^{11})	CdTe	–	–	–	–	1.3	Watts and Holton (1967)
	CdS	–	–	–	–	1.3	Watts and Holton (1967)
Tm^{2+} (f^{13})	CdTe	2.647	–	–	–	1.3	Watts and Holton (1967)
Yb^{3+} (f^{13})	ZnTe	3.182	–	–	$A^{171} = 860$ $A^{173} = 238$	1.3	Watts (1966)
	CdTe	3.117	–	–	$A^{171} = 860$ $A^{173} = 232$	1.3	Watts (1966)

[a]) The parameters b_4, b_6 correspond to a different notation often used to calculate positions of ESR lines, especially with rare earth ions (for details, see Low and Offenbacher, 1966).

Eu^{2+}, Gd^{3+} $(4f^1)$, $^8S_{\frac{7}{2}}$ ground state. Eu^{2+} and Gd^{3+} have been observed in some cadmium compounds and Gd^{3+} in ZnO. Gd^{3+} in ZnO is substitutional for zinc, the charge compensation is at least some lattice points away. The existence of a distortion of the lattice in the vicinity has been concluded from an anisotropy of the resonances. A weak influence of covalency indicates that more than one perturbation is responsible for the ground state splitting of the $^8S_{\frac{7}{2}}$ ion in axial symmetry.

Dy^{3+} (f^9). Axial Dy^{3+} resonances have been detected in hexagonal ZnS and CdS. It seems possible that the centres are isolated substitutional Dy^{3+} ions. The spectra show the influence of a trigonal field superimposed on a tetrahedral one.

Er^{3+} and Ho^{2+} $(4f^{11})$. Spectra of Er^{3+} have been observed in ZnS, ZnSe, ZnTe, CdS and CdTe in three different sites. For the ESR spectra of Er^{3+} in CdTe, as well as for the isoelectronic Ho^{2+}, a definite analysis has not yet been given.

TABLE 5.5
ESR parameters of rare earth impurities in hexagonal II–VI compounds

Ion	Crystal	$g_\parallel$	$g_\perp$	b_2^1 $(10^{-4}$ $cm^{-1})$	b_1^2 $(10^{-4}$ $cm^{-1})$	b_2^2 $(10^{-4}$ $cm^{-1})$	b_1^3 $(10^{-4}$ $cm^{-1})$	$A_\parallel$ $(10^{-4}\,cm^{-1})$	$A_\perp$ $(10^{-4}\,cm^{-1})$	T (K)	References
Nd^{3+} (f^3)	CdS	0.43	3.409	–	–	–	–	$A^{143}=342$	$B^{143}=213$	1.3	Morigaki (1963c)
	CdSe	0.7	3.45	–	–	–	–	–	–	4.2	Title (1967)
	CdTe	3.91	1.25	–	–	–	–	–	–	4.2	Title (1967)
Nd^{2+} (f^4)	CdS	4.30	0	–	–	–	–	–	–	1.3	Morigaki (1963c)
	CdSe	5.17	0	–	–	–	–	–	–	1.3	Morigaki (1963c)
Eu^{2+} (f^7)	CdS	1.992	1.992	−336.6	−11.6	0.69	–	$A^{151}=22.5$ $A^{153}=10.0$	–	300	Dorain (1960)
	CdSe	1.994	1.994	+239.7	−2.75	0.24	11.7	$A^{151}=23.2$ $A^{153}=10.2$ $A^{155}=-2.24$ $A^{157}=-2.98$	$B^{155}\leqq0.25$ $B^{157}\leqq0.3$	77	Title (1964a)
Gd^{3+} (f^7)	ZnO	1.987	1.978	+916.7	+17.0	+0.5	–	–	–	300	Hausmann (1969c)
	CdSe	1.985	1.985	+805.5	+6.91	+0.36	17.5	–	–	77	Title (1965)
Dy^{3+} (f^9)	ZnS	10.02	4.109	–	–	–	–	–	–	1.3	Watts and Holton (1967)
	CdS	8.378	5.99	–	–	–	–	–	–	1.3	Watts and Holton (1967)
Er^{3+} (f^{11})	CdS	11.42	1.68	–	–	–	–	–	–	1.3	Watts and Holton (1967)
Yb^{3+} (f^{13})	ZnO	1.311	4.421	–	–	–	–	$A^{171}=328$ $A^{173}=90.7$	$B^{171}=1165$ $B^{173}=323$	15	Hausmann and Schreiber (1970)
	ZnS	1.242	4.40	–	–	–	–	$A^{171}=330$ $A^{173}=90$	$B^{171}=1174$ $B^{173}=320$	1.3	Watts (1966)
	ZnTe	1.14	4.49	–	–	–	–	–	–	4.2	Watts (1966)
	CdS	1.1856	4.439	–	–	–	–	$A^{171}=340.5$	$B^{171}=1166$	1.3	Watts (1966)

Tm^{2+} ($4f^{13}$). An isotropic resonance due to Tm^{2+} has been found in CdTe. The ion is assumed to be interstitial at the same interstice occupied by Yb^{3+}.

Yb^{3+} ($4f^{13}$). Spectra of isolated substitutional Yb^{3+} ions have been detected in ZnTe, CdTe and hexagonal ZnO, ZnS, ZnTe and CdS. In ZnTe doped with Yb and P, Yb can be found in a place of trigonal symmetry. Here Yb is at a Zn site with P occupying one of the four neighbouring Te sites. In CdTe doped with a noble metal, two types of rare earth, noble metal associates occur. In each type, the trivalent rare earth is interstitial at the site whose nearest neighbours are four Cd atoms along the $\langle 111 \rangle$ axes and six next nearest neighbour Te atoms along the $\langle 100 \rangle$ axes.

Summarizing, it can be said that still more measurements, resonance as well as luminescence, have to be performed to establish definite models of these centres and their surroundings.

5.4. Mobile electrons

Mobile conduction electrons in a solid belong to the most elementary sources of paramagnetism. They give rise to the conductivity of metals, and electrons or holes moving in the conduction or valence band and/or shallow donor bands cause the conductivity of n or p type semiconductors.

As a conduction electron moving in a solid is only well defined in momentum (k) space, but not in coordinate (r) space, the discussion of ESR of mobile carriers is somewhat outside the general line of this book, which is dealing with paramagnetic defects, that means with spins which are localized in coordinate space. Still it seems to be quite reasonable to discuss the ESR results of mobile carriers together with the other topics of this book, for molecular orbital theory shows that a spin entirely localized in (r) space is an idealization and some delocalization of the unpaired electron wave function is admitted. On the other hand, some localization of mobile carriers by means of interaction with lattice vibrations has to be considered as well.

So far only ESR of mobile electrons has been observed for several of the II–VI compounds, whereas ESR of mobile holes has not yet been detected. The mobile electrons in a semiconductor are thought to move either in the conduction band of the crystal and/or in shallow donor bands. These donor bands are formed in the following way: Substitutional impurities that act as donors in the II–VI compounds consist either of elements of column III

(Al, In, Ga, Tl) of the periodic table or of column VII (Cl, Br, I) elements. In both cases the impurity has one more valence electron than the host atom for which it substitutes. These extra electrons serve as donor electrons. (Although anion vacancies may act as donors, up to now there is no definite evidence for such an example in the II–VI compounds.)

The donor electrons are very weakly bound to the donor core and their orbit is therefore highly delocalized. At a certain value of the donor concentration the wave functions of the electrons start to overlap. The donor electrons may then move from donor to donor and are no longer localized at the impurity site. With further increasing concentration of donors, the overlapping wave functions form a donor band which finally may even touch the conduction band. The donor electrons may now move in these bands; however, the mobility is less in the donor band than in the conduction band.

It is known that an electron moving in a periodic potential of a crystal behaves as if it has an effective mass m^*, which is different from that of the free electron m. An expression for the effective mass m^* and the motion of electrons in semiconductors can be most conveniently obtained by means of the $k \cdot p$ perturbation theory (Roth, 1960). In this method, an appropriate wave function for the donor electrons is chosen as a product of a Bloch type conduction band function and a hydrogen-like wave function. Taking properly into account spin–orbit interaction, one gets the result that the g factor of a mobile electron also differs from the free spin value. The calculations generally are very difficult for indirect energy gap semiconductors as it requires exact knowledge of the energetical positions of all bands.

By means of spin–orbit interaction, the electron spin of a mobile electron can notice collisions with the lattice phonons or with lattice defects, which in a real crystal are responsible for a finite conductivity. The widths of the ESR lines are in some way related to the collision frequency of the carriers (Elliot, 1954). Therefore the line width is a quantity which is characteristic for the crystal as a whole and not for a particular type of lattice defect.

In no case, hyperfine interaction of the unpaired electrons with the nuclear spin of the donor has been detected. At liquid helium temperatures, the donor core normally should have retrapped most of the unpaired electrons thus making ESR of the neutral donor centre observable. Only at higher temperatures, thermal ionization can create conduction electrons either across the energy gap (intrinsic conductivity) or from shallow donor impurities. (Here we do not consider optical excitation of photoelectrons.) Therefore we need an explanation for the fact that in a semiconductor even at very low temperatures the donor centres are still ionized and mobile electrons do exist in the crystal.

If we discuss the bound states of a donor electron in terms of a hydrogen-like model, the result is that the energy of the ground state of the electron, i.e. its energetical separation from the bottom of the conduction band, is very much reduced below the energy of the free hydrogen atom due to the effective mass and the dielectric constant of the host lattice. Parallel to the strong reduction of the donor ground state energy, one may expect an increase of the Bohr's radius of the excess electron over that of the free hydrogen atom. This means that the unpaired donor electron is spread over many unit cells of the crystal. We now understand that the concept of the isolated donor is only meaningful in the limit of low donor concentration. Only then overlap of the wave functions of the individual donors can be neglected.

So far, ESR of isolated shallow donor centres in semiconductors has only been observed for silicon and germanium. Only these crystals have been produced in such ultrapure form that overlap of the donor wave functions could be neglected. In all other semiconductors, the excess electrons can still move from donor to donor due to the high impurity rate of the material.

The comparatively low purity of other semiconductors than Si and Ge permits only observation of ESR of mobile electrons. As the donor electrons are not loalized at a donor core, the hyperfine interaction with the donor nuclei is smeared out and averaged to zero. The absence of any hyperfine structure to identify the donor centre and the appearence of a single line with a g value less than 2 are features common to the resonances of all group III or VII substitutional donors in the II–VI compounds. Mostly identification of the centre responsible for the resonance has been made from knowledge of the donor impurity added during specimen preparation.

In table 5.6, a summary of experimental data for mobile electrons in II–VI semiconductors is given. Zeeman effect studies of exciton lines also yield a g value of conduction electrons. A Mott–Wannier type exciton is a weakly bound electron–hole pair whose bound energy states can be described by a similar hydrogen-like model as the donor centres. Then we only have to replace m^* by the reduced effective mass of the electron–hole pair. The exciton energy states show up in the optical spectra of a semiconductor as a Rydberg-like series of sharp lines appearing on the low energy side of the band–band absorption edge. Although the g values obtained by Zeeman effect studies are less accurate than those of ESR, they are included too. The

TABLE 5.6

ESR and Zeeman effect results on donor centres in II–VI compounds

Host lattice	Doping	T (K)	g	$g_\parallel$	$g_\perp$	g	$g_\parallel$	$g_\perp$	References
							Zeeman effect		
ZnS (cubic)	I	77	1.8823	–	–	–	–	–	Müller and Schneider (1963)
	Cl	77	1.8835	–	–	–	–	–	Müller and Schneider (1963)
	Al	77	1.8849	–	–	–	–	–	Müller and Schneider (1963)
ZnSe (cubic)	–	77	1.14	–	–	–	–	–	Schneider and Räuber (1966)
ZnO (hex.)	Zn, In	77, 300	–	1.956	1.955	1.95	–	–	Müller and Schneider (1963)
									Reynolds et al. (1964)
ZnS (hex.)	Ga, Al	77, 300	–	1.9557	1.9552	–	–	–	Müller and Schneider (1963)
	Cl	77	–	1.8933	1.8860	–	1.9	2.2	Müller and Schneider (1963)
CdS (hex.)	Cl	4.2	–	1.792	1.775	–	1.78	1.72	Miklosz and Wheeler (1967)
	Ga	77	–	1.78	1.72	–	–	–	Dieleman (1962)
									Lambe and Kikuchi (1958)
									Hopfield and Thomas (1961)
CdSe (hex.)	–	1.8	–	–	–	–	0.6	0.51	–
CdO (NaCl)	–	77	1.806	–	–	–	–	–	Müller and Schneider (1963)

reason is that sometimes (CdS) the magneto-optical method for determining g still can be used if ESR lines are already broadened beyond detection.

In the case of the hexagonal semiconductors, the axial crystal field symmetry is reflected by a weak anisotropy of the g factor. In n type material, the donor resonances can be observed in the dark, whereas in insulating crystals mostly photoexcitation is necessary.

ESR of mobile electrons has also been observed in cubic ZnS/ZnSe mixed crystals (Schneider et al., 1968). It was found that the g factor of the conduction electrons shifts continuously as the concentration of the components of the binary system is changed. This behaviour can be seen in fig. 5.5. This feature can well be explained by the model of Müller and Schneider (1963). Since the electron wave function is highly delocalized, the electron orbit extending over several lattice spacings, the excess electron cannot resolve the statistical fluctuations of the microscopic lattice symmetry and an average g value is measured.

Summarizing the observations obtained by ESR measurements of donor electrons in II–VI semiconductors, some features can be noticed which hold for all resonances so far detected. Substitutional donors give rise to mobile electrons in donor impurity bands and/or the conduction band. No hyperfine interaction with the neutral donor has yet been resolved. The g values seem to be characteristic of the intrinsic band properties of the compound. Mostly the resonances can be enhanced by optical excitation or lowering of the temperature to trap the mobile electrons in the impurity band. The observation of localized donor resonances would be useful in mapping the spatial extent of the wave function of the donor and for the study of trap-

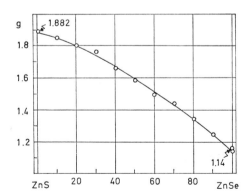

Fig. 5.5. The g value of mobile electrons in ZnS : ZnSe mixed crystals as a function of the ZnSe content (in %) ($v = 9.1$ GHz, $T = 77$ K). (After Schneider et al., 1968.)

ping at donor sites during processes of photoconductivity or photoluminescence. However, it is unlikely that this can be done until the compounds can
be purified to such an extent that overlapping of the donor wave functions
does not take place any longer.

5.5. Acceptor centres

Besides the many reported resonances of donor centres in the II–VI
compounds, only rather few resonances have been identified with acceptor
centres. Analogously to donor centres, acceptor centres are formed if a
cation of group I (Li, Na) replaces a group II element or if a group VI
element is substituted by a group V element (P, As, Sb) at the anion site.
Here the substitute has one electron less and can act as an acceptor, for
instance Cu.

The only isolated acceptor centre of group I elements which has been
observed up to now is the lithium acceptor centre in ZnO (Schirmer, 1968).
The conductivity of n type ZnO can be highly suppressed by doping the
crystal with lithium. At low temperatures, a deep acceptor centre has been
identified by ESR after irradiation with ultraviolet light. The resonances

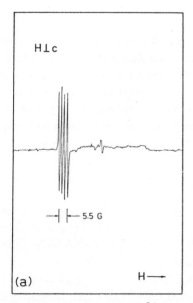
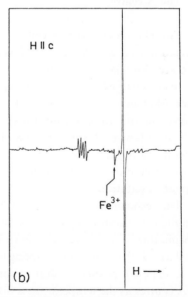

Fig. 5.6. ESR spectrum of an Li^{2+} centre in a ZnO crystal after UV irradiation at
77 K ($\nu = 9.091$ GHz). The transitions occur at $g_\perp = 2.0253$ and $g_{||} = 2.0028$. The weaker
lines arise from nonaxial centres. (After Schneider and Schirmer, 1963.)

show axial symmetry around the c axis of the hexagonal crystals. The spectrum can be seen in fig. 5.6. For an orientation of the magnetic field normal to c, it consists of a well-resolved quartet pattern because of an interaction of the unpaired spin $I = \frac{3}{2}$ of the Li^7 nucleus; HFS is not resolved for H parallel to c.

The centres become paramagnetic by trapping an excited hole at one of the four oxygen ligands. Two types of centres have to be distinguished in case the hole is trapped at the oxygen ligand parallel to the c axis or at one of the three normal to c. The trapping of the hole in the different ligands results in a modified symmetry of the electric crystal field giving rise to different g values. If the hole is trapped at one of the three equivalent non-axial ligands, it is in a thermally excited state, the separation in energy from the ground state being 0.015 eV. Optical excitation may disturb the thermal equilibrium between the two types of centres.

Studies of the two isotopes of lithium, Li^6, having nuclear spin $I = 1$, and Li^7, revealed that the g tensor depends on the mass of the isotopes. The resultant g shift of about 2×10^{-5} can be explained as being due to the zero point vibrations of the impurity ion.

5.6. The A centre

For a complex acceptor centre, named 'A centre', resonances have been detected in ZnS and ZnTe as well as in mixed ZnS/ZnSe crystals. These defects are dominant impurity centres formed by association of a double negatively charged zinc vacancy with one nearest anion site replaced by either a halogen ion Cl^-, Br^- or I^-, or with a trivalent Al or Ga ion on a nearest zinc site. In both cases the centre lacks a positive charge with respect to the host lattice. Resonance data of A centres in the II–VI compounds are summarized in table 5.7. Normally the A centre is diamagnetic and has to be converted into a paramagnetic state by trapping a hole which is produced by photo excitation.

The resonances have some features in common. The g values are larger than the free electron value. This is the reason for identifying the resonances as being due to a hole centre. Generally one takes the g shift to be proportional to $-\lambda$, the spin–orbit coupling constant. As a hole has a negative λ, one has to expect a positive g shift. Another fact is the change of symmetry of the A centres as the temperature is lowered from 77 to 4.2 K, indicating that we are dealing with associated centres. At 77 K, the centres have axial symmetry, whereas at liquid helium temperature the symmetry is orthorhombic.

TABLE 5.7

ESR parameters of the A centres in the II–VI compounds

Host	Centre	g_1	g_2	g_3	References
ZnS (hex.)	V_{Zn}–Cl	$g_{\parallel} = 2.059$	$g_{\perp} = 2.0287$		Schneider et al. (1963a)
	V_{Zn}–Ga	2.004	2.0534	2.0587	Otomo et al. (1963)
ZnS (cubic)	V_{Zn}–Cl	2.0027	2.0502	2.0565	Schneider et al. (1965)
	V_{Zn}–Br	2.0029	2.0537	2.0569	Holton et al. (1965)
	V_{Zn}–I	–	2.048	–	Holton et al. (1965)
	V_{Zn}–Al	2.0030	2.0513	2.0560	Räuber and Schneider (1962b)
	V_{Zn}–Ga	2.0025	2.0509	2.0557	Holton et al. (1965)
ZnSe	V_{Zn}–Cl	1.9597	2.1612	2.2449	Holton et al. (1965)
ZnTe	V_{Zn}–Al	2.045	2.088	2.091	Title et al. (1964)
ZnS : Se	V_{Zn}–Cl	1.9708	2.1742	2.1941	Schneider et al. (1967)
	V_{Zn}–Br	1.9708	2.1825	2.1922	Schneider et al. (1967)

Measurements on samples doped with different donors indicated that the A centre has to be assigned to resonances of a zinc vacancy and a neighbouring substitutional donor. This model has now been verified by the observation of a weak ligand hyperfine interaction between the hole and the donor nucleus (Schneider et al., 1965). A more detailed model of the electronic structure of an A centre is shown in fig. 5.7. Four single pair orbitals are directed versus the zinc vacancy. Most time, the hole will be localized in the sulphur single pair orbitals as the electrons in the halogen orbital are more tightly bound. Taking into account overlap of these orbitals and a Jahn–Teller distortion which tends to remove the degeneracy of the energy states of the centre further, one obtains the following result: The unpaired hole is dominantly localized at only *one* sulphur site. At higher temperatures, thermally activated hopping of the hole from one of the three sulphur sites to another is observed. The activation energy involved in this process

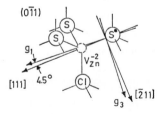

Fig. 5.7. A model of the A centre in ZnS. The A centre shown is an association of a double negatively charged zinc vacancy V_{Zn}^{2-} and a group VII impurity; Cl occupying one of the neighbouring S sites. The hole trapped after optical excitation is shown as * on one of the sulphur sites. The orthorhombic axes g_1 and g_3 are shown; g_2 is along the [011] direction. (After Schneider et al., 1963.)

amounts to 0.057 eV for the A centre in ZnS. In case of a group III impurity, the unpaired spin is permanently locked to that sulphur site which is around the Zn vacancy and has the greatest distance from the donor.

The spectrum of the A centre can be characterized by a spin $S = \frac{1}{2}$ and an orthorhombic g tensor. The unpaired spin is localized in the nonbonding single pair sulphur orbital pointing towards the Zn vacancy, whereas the three bonding orbitals are formed by the six valence electrons of sulphur and its three nearest zinc ligands. The deviations of the g tensor components of the A centres from the free spin value is understood as a result of spin–orbit interaction which admixes mainly bonding states to the unpaired electron wave functions. From measurements on A centres in ZnS crystals containing some percents of ZnSe, it has been concluded from the observation of HFS with Se[77] nuclei that the single pair orbital has mainly s character and hence the bonding orbitals must have dominantly sp² character.

The paramagnetic A centre has trapped a hole. Subsequently, a conduction electron may recombine with the trapped hole arising in a characteristic luminescence in the visible region, named the 'self activated' emission. The self activated blue emission of ZnS is thought to occur if the conduction electron is caught in the antibonding orbitals, corresponding to a transition from an excited state of the nonionized A centre into its diamagnetic ground state. These ideas are strongly supported by experiments which relate the self activated luminescence to the intensity of the ESR spectra of the A centres (Prener and Weil, 1959).

5.7. The F centre in ZnS

Another type of a diamagnetic centre which can be converted into a paramagnetic state by trapping an electron is the F centre which has been observed in ZnS (Schneider and Räuber, 1967) and ZnO (Hausmann, 1970). The centre is formed if an electron is trapped by a double positively charged anion vacancy. Fast neutron irradiation or annealing of the crystals in liquid zinc as well as mechanical damage, such as crushing of the crystal, are possibilities for producing F centres.

The wave function of the F centre is expected to have mainly s character (Zn^+). Thus the spectrum has an isotropic g factor of $g = 2.0034$ in ZnS. An additional weak hyperfine structure is found, arising from interaction with the unpaired spin of four equivalent nuclei of the 4.1 % abundant Zn^{67} isotopes ($I = \frac{5}{2}$) in the first zinc shell around the vacancy. The HFS is slightly anisotropic as referred to the $\langle 111 \rangle$ axes, with

$$A_\parallel = 28.0 \times 10^{-4} \text{ cm}^{-1}, \qquad A_\perp = 23.3 \times 10^{-4} \text{ cm}^{-1}.$$

Fig. 5.8 shows the spectrum for the F centre in ZnS. This centre is the first intrinsic lattice defect of the II–VI semiconductors of which ESR has been detected. Optical measurements on the F centres have been reported by Leutwein and Räuber (1967).

A defect related to an F centre is a centre in which the central sulphur vacancy is occupied by a diamagnetic F^- ion. ESR of this defect has been observed in ZnS by Kasai (1965). Hyperfine structure interaction of the central fluorine nucleus with four equivalent nearest neighbour and 12 equivalent next nearest neighbour zinc nuclei has been detected in this case. Evidently, the unpaired electron wave function is expelled from the central anion site as is indicated by the values of the ESR parameters. Fluorine does not form a donor state in the II–VI compounds, in contrast to the other halogen ions.

ESR of F centre Cu^+ complexes and associates has been reported by Dieleman et al. (1964).

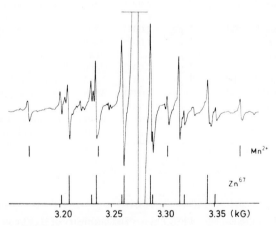

Fig. 5.8. Zn^{67} hyperfine interaction of the unpaired F centre electron in ZnS, for $H \parallel$ [111] ($\nu = 9.186$ GHz, $T = 300$ K). Under this orientation two types of nearest neighbour zinc ligands can be distinguished, having statistical weights of 1 : 3. The strong signal in the centre arises from those zinc isotopes which lack nuclear spin. Additional signals are seen which may in part arise from forbidden hyperfine transitions or from hexagonal domains of the crystal. (After Schneider and Räuber, 1967.)

5.8. S state centres

A group of paramagnetic centres with S ground states has been observed in ZnO, ZnS, ZnSe and ZnTe crystals. Räuber and Schneider (1966) have detected Ga^{2+}, In^{2+} and Tl^{2+} impurities; trivalent ions of silicon and germanium and of Sn^{3+} and Pb^{3+} have also been observed. The centres seem to be located substitutionally at zinc sites. A common feature of these centres is a very large hyperfine interaction with the nuclear spin of the impurity ion. The ESR parameters of the S state ions are summarized in table 5.8. The group III impurities Ga, In and Tl do not seem to form shallow donor states, in contrast to Al. All the S state centres are photosensitive.

TABLE 5.8
ESR parameters of $^2S_{1/2}$ state ions in ZnS and ZnTe

Host lattice	Impurity ion		Nuclear spin	g	$\|A\|$ $(10^{-4}$ $cm^{-1})$	References
ZnO	Sn^{117}	$5s^1$	$^1/_2$	1.9877	3181	Hausmann and Schreiber (1971)
	Sn^{119}		$^1/_2$	1.9877	3327	
ZnS	Si^{29}	$3s^1$	$^1/_2$	2.0047	654	Sugibuchi and Mita (1966)
	Ga^{69}	$4s^1$	$^3/_2$	1.9974	2025	Räuber and Schneider (1966)
	Ga^{71}		$^3/_2$	1.9974	2574	
	Ge^{73}	$4s^1$	$^9/_2$	2.0086	305	Sugibuchi and Mita (1966)
	In^{115}	$5s^1$	$^9/_2$	1.9930	3123	Räuber and Schneider (1966)
	Sn^{117}	$5s^1$	$^1/_2$	2.0075	5207	Sugibuchi (1967)
	Sn^{119}		$^1/_2$	2.0075	5448	
	Tl^{203}	$6s^1$	$^1/_2$	2.0095	23640	Räuber and Schneider (1966)
	Tl^{205}		$^1/_2$	2.0095	23860	
ZnSe	Ge^{73}	$4s^1$	$^9/_2$	2.043	260	Suto and Aoki (1969)
	Sn^{117}	$5s^1$	$^1/_2$	2.0251	4556	Holton and Watts (1969)
	Sn^{119}		$^1/_2$	2.0251	4767	
	Pb^{207}	$6s^1$	$^1/_2$	2.073	6249	Holton and Watts (1969)
ZnTe	Pb^{207}		$^1/_2$	2.167	5230	Suto and Aoki (1967)

References

Aven, M. and J. S. Prener, eds., 1967, Physics and chemistry of II–VI compounds (North-Holland, Amsterdam).
Bleaney, B., 1959, Proc. Phys. Soc. London **73**, 939.
De Kinder, R. E., Jr., 1965, Bull. Am. Phys. Soc. **10**, 173.
De Wit, M. and A. R. Reinberg, 1967, Phys. Rev. **163**, 261.
De Wit, M., A. R. Reinberg, W. C. Holton and T. L. Estle, 1965, Bull. Am. Phys. Soc. **10**, 329.

Dieleman, J., 1962, in: J. Smidt, ed., Magnetic and electric resonance and relaxation (North-Holland, Amsterdam) 409.

Dieleman, J., R. S. Title and W. V. Smith, 1962, Phys. Letters **1**, 334.

Dieleman, J., S. H. de Bruin, C. Z. van Doorn and J. H. Haanstra, 1964, Philips Res. Rept. **19**, 311.

Dorain, P. B., 1960, Phys. Rev. **120**, 1190.

Dziesiaty, J. and R. Böttcher, 1968, Phys. Stat. Sol. **26**, K 21.

Elliot, R. J., 1954, Phys. Rev. **96**, 226.

Estle, T. L. and W. C. Holton, 1965, Bull. Am. Phys. Soc. **10**, 57.

Estle, T. L. and W. C. Holton, 1966, Phys. Rev. **150**, 159.

Hausmann, A., 1968, Solid State Commun. **6**, 457.

Hausmann, A., 1969a, J. Phys. Soc. Japan **26**, 91.

Hausmann, A., 1969b, Phys. Stat. Sol. **31**, K 131.

Hausmann, A., 1969c, Solid State Commun. **7**, 579.

Hausmann, A., 1970, Z. Physik **237**, 86.

Hausmann, A. and H. Huppertz, 1968, J. Phys. Chem. Solids **29**, 1369.

Hausmann, A. and P. Schreiber, 1969, Solid State Commun. **7**, 631.

Hausmann, A. and P. Schreiber, 1970, Solid State Commun. **8**, 1103.

Hausmann, A. and P. Schreiber, 1971, Z. Physik **245**, 184.

Hennig, J. C. M., H. van de Boom and J. Dieleman, 1969, Philips Res. Rept. **21**, 16.

Henzel, J. C., 1964, Bull. Am. Phys. Soc. **9**, 244.

Holton, W. C. and R. K. Watts, 1969, Phys. Rev. **51**, 1615.

Holton, W. C., J. Schneider and T. L. Estle, 1964, Phys. Rev. **133**, 1638.

Holton, W. C., M. de Wit and T. L. Estle, 1965, Bull. Am. Phys. Soc. **9**, 249.

Hopfield, J. J. and D. G. Thomas, 1961, Phys. Rev. **122**, 35.

Hoshina, T., 1966, J. Phys. Soc. Japan **21**, 1608.

Kasai, P. H., 1965, J. Chem. Phys. **43**, 4143.

Lambe, J. and C. Kikuchi, 1958, J. Phys. Chem. Solids **9**, 492.

Lambe, J. and C. Kikuchi, 1960, Phys. Rev. **119**, 1256.

Lambe, J., J. Baker and C. Kikuchi, 1959, Phys. Rev. Letters **3**, 270.

Leutwein, K. and A. Räuber, 1967, Solid State Commun. **5**, 783.

Low, W. and E. L. Offenbacher, 1966, Solid State Phys. **17**.

Ludwig, G. W. and M. R. Lorenz, 1963, Phys. Rev. **131**, 601.

Ludwig, G. W. and H. H. Woodbury, 1962, Solid State Phys. **13**, 299.

Matamura, O., 1959, J. Phys. Soc. Japan **14**, 108.

Miklosz, J. C. and R. G. Wheeler, 1967, Phys. Rev. **153**, 913.

Morigaki, K., 1963a, J. Phys. Soc. Japan **18**, 733.

Morigaki, K., 1963b, J. Phys. Soc. Japan **18**, 1558.

Morigaki, K., 1963c, J. Phys. Soc. Japan **18**, 1636.

Morigaki, K., 1964a, J. Phys. Soc. Japan **19**, 1240.

Morigaki, K., 1964b, J. Phys. Soc. Japan **19**, 1485.

Morigaki, K. and T. Hoshina, 1965, Phys. Letters **17**, 85.

Müller, K. A. and J. Schneider, 1963, Phys. Letters **4**, 288.

Otomo, Y., H. Kusumoto and P. H. Kasai, 1963, Phys. Letters **4**, 228.

Owen, J. and K. W. Stevens, 1953, Nature **171**, 836.

Prener, J. S. and D. J. Weil, 1959, J. Electrochem. Soc. **106**, 409.

Rai, R., J. Y. Savard and B. Tousignant, 1967, Phys. Letters **25**, 443.

Räuber, A. and J. Schneider, 1962, Z. Naturforsch. **17a**, 266.

Räuber, A. and J. Schneider, 1962b, Phys. Letters **3**, 230.

Räuber, A. and J. Schneider, 1966, Phys. Stat. Sol. **18**, 125.

Reynolds, D. C., C. W. Litton and T. C. Collins, 1965, Phys. Rev. **140**, A1726.

Roth, L. M., 1960, Phys. Rev. **118**, 1534.

Schirmer, O., 1968, J. Phys. Chem. Solids **29**, 1407.
Schneider, J. and A. Räuber, 1966, Phys. Letters **21**, 380.
Schneider, J. and A. Räuber, 1967, Solid State Commun. **5**, 779.
Schneider, J., W. C. Holton, T. L. Estle and A. Räuber, 1963a, Phys. Letters **5**, 312.
Schneider, J., S. R. Sircar and A. Räuber, 1963b, Z. Naturforsch. **18a**, 980.
Schneider, J., A. Räuber, N. Dischler, T. L. Estle and W. C. Holton, 1965, J. Chem. Phys. **42**, 1839.
Schneider, J., B. Dischler and A. Räuber, 1967, in: Proc. Intern. Conf. on II–VI semiconducting compounds (Benjamin, New York) 55.
Schneider, J., B. Dischler and A. Räuber, 1968, J. Phys. Chem. Solids **29**, 451.
Sturge, M. D., 1967, Solid State Phys. **20**.
Sugibuchi, K., 1967, Phys. Rev. **153**, 404.
Sugibuchi, K. and Y. Mita, 1966, Phys. Rev. **147**, 355.
Suto, K. and M. Aoki, 1967, J. Phys. Soc. Japan **22**, 1307, 1517.
Suto, K. and M. Aoki, 1969, J. Phys. Soc. Japan **26**, 287.
Title, R. S., 1959, Phys. Rev. Letters **3**, 273.
Title, R. S., 1964a, Phys. Rev. **133**, A 198.
Title, R. S., 1964b, Phys. Rev. **133**, A 1613.
Title, R. S., 1965, Phys. Rev. **138**, A 631.
Title, R. S., 1967, in: M. Aven and J. S. Prener, eds., Physics and chemistry of II–VI compounds (North-Holland, Amsterdam).
Title, R. S., G. Mandel and F. F. Morehead, 1964, Phys. Rev. **136**, A 300.
Van Wieringen, J. S., 1955, Discussions Faraday Soc. **19**, 118.
Watanabe, H., 1964, J. Phys. Chem. Solids **25**, 1471.
Watts, R. K., 1966, Solid State Commun. **4**, 549.
Watts, R. K., 1969, Phys. Rev. **188**, 568.
Watts, R. K. and W. C. Holton, 1967, in: Proc. Intern. Conf. on II–VI semiconducting compounds (Benjamin, New York) 1396.
Woodbury, H. H. and G. W. Ludwig, 1961, Bull. Am. Phys. Soc. **6**, 118.

6 | III – V SEMICONDUCTING COMPOUNDS

In only a few of the III–V compounds spin resonance has so far been observed, namely in GaP, GaAs, InAs and InSb. Up to now, the great difficulty still is the preparation of single crystals of high purity and controlled doping. At concentrations exceeding some 5×10^{15} impurity atoms per cm^3, the ESR lines are generally wide and difficult to detect. Most of the prepared crystals are degenerate even below room temperature because of uncontrolled impurities which are the reason that conduction electrons are present.

ESR of isolated neutral donor or acceptor centres in semiconductors is a rare event which so far has only been detected in silicon, germanium and SiC. The concept of an isolated donor in any case will only be meaningful in the limit of low donor concentration, when overlap of the individual electron wave functions can be neglected. (The case of overlap of the wave functions has in some detail been discussed in section 5.4.)

6.1. Mobile electrons

Most defects observed in the III–V semiconductors are due to conduction electrons moving in impurity bands or the conduction band. One of the III–V compounds, InSb, is a very suitable substance for studying the behaviour of mobile electrons. InSb is a cubic direct gap semiconductor. It has spherical energy bands and the smallest energetical separation occurs at $k = 0$. For this case, the effective mass of the electrons can be calculated from the Hamiltonian which acts on the Bloch function of the electron and takes properly into account spin–orbit interaction. Also the g factor of the electron moving in a periodic potential can be calculated for this case in a rather simple manner.

177

In InSb, the electrons move in an s type conduction band separated
from the p type valence band by only 0.2 eV. At $k = 0$, the Bloch function
resembles an atomic 5s function of In^{2+}, the valence band at $k = 0$ has
5p character as for Sb^{2-}. The spin–orbit interaction causes the degenerate
valence band to be further split. Roth et al. (1959) calculated the effective
moment μ of the mobile electrons as

$$\mu = \mu_B \left[1 - \frac{(m_e/m^* - 1)\ \lambda}{3E_g + \lambda} \right],$$

where λ is a spin–orbit constant giving the splitting of the valence band
and E_g is the energy gap. The g factor, equal to $2\mu/\mu_B$ as predicted from
this equation, is about 50. Bemski (1960b) has observed the spin resonance
of conduction band electrons over a range of concentrations from 2×10^{14}
to 3×10^{15} electrons per cm^3 in InSb. The above relation was excellently
verified by experiment, g ranging from 48.8 to 50.7 and varying according
to the position of the Fermi level. In this case, ESR has provided a possibility for measuring the departure of the conduction band from a parabolic
shape (Kane, 1957).

Konopka (1967) has observed resonances of mobile electrons in InAs
at 4.2 K. He finds a linear variation of the g value with temperature, which
can be understood by variation of the Fermi level due to the condensation
of part of the free carriers on the impurity levels.

Irradiated n type InSb has been investigated by Kaplan and Guéron
(1965). Irradiation created electron–hole pairs which give part of their
energy to the free carriers. This gives rise to an increase of several degrees
in the kinetic temperature of the carriers, which is measured by the variation
of the g factor.

ESR results for mobile electrons in III–V semiconductors so far obtained are summarized in table 6.1.

TABLE 6.1

g factors of mobile electrons in III–V compounds

Host lattice	T (K)	g	References
GaP	77	1.998	Title (1967)
GaAs	4.2	0.52	Duncan and Schneider (1963)
InAs	4.2	−14.7	Konopka (1967)
InSb	4.2	≈ 50	Bemski (1960a)

6.2. Other centres

Compared with the II–VI semiconductors, there is only a very small number of ESR results of paramagnetic centres which have been published for the III–V compounds. The reason has been pointed out earlier. Data so far obtained are collected in table 6.2.

TABLE 6.2
Resonance parameters of centres in group III–V semiconductors

Lattice	Centre	g	$g_{\parallel}$	$g_{\perp}$	a $(10^{-4}$ $cm^{-1})$	A $(10^{-4}$ $cm^{-1})$	References
GaP	Mn^{55}	2.002	–	–	–	$\pm\,55$	Woodbury and Ludwig (1961)
GaP	Fe^{3+} $(3d^5)$	2.025	–	–	390	–	
GaAs	Fe^{3+}	2.045	–	–	340	–	De Wit and Estle (1962,1963)
GaAs	$Ni\,(3d^7)$	2.106	–	–	–	–	De Wit and Estle (1962)
GaAs (stressed)	Cd	–	3.4	6.7	–	–	Fedotov et al. (1968)
GaAs	Cd, Zn	–	–	–	–		Title (1963)

Three different 3d ion centres have been detected in GaP and GaAs. Woodbury and Ludwig (1961) have studied Fe and Mn impurities in GaP. The iron spectrum shows five fine structure transitions and $g \approx 2$, indicating that $S = \frac{5}{2}$. It was believed that Fe substitutes for Ga and promotes one 3d shell electron to the valence shell, completing the tetrahedral bonding. Possibly the manganese ion has also $S = \frac{5}{2}$. However, only the central transition is observed. The other transitions might be broadened beyond detection.

Resonances of Fe and Ni have been detected by De Wit and Estle (1963) in GaAs. For iron, a typical five line spectrum is found which can be explained by $S = \frac{5}{2}$ and an isotropic g value close to 2. The iron ion is incorporated in a cubic crystal field environment. For Ni, only one broad line is found and attributed to a $3d^7$ Ni configuration in a tetrahedral symmetry. The ions are supposed to substitute for Ga.

Some results have been reported on shallow acceptor centres. In GaAs, the degeneracy of the valence band prevents direct observation of the ESR of these centres. The degeneracy can be lifted by uniaxial compression. Fedotov et al. (1968) have observed resonances of shallow Cd acceptors in mechanically stressed single crystals of GaAs. The g values agree with those expected for free holes. Title (1963) has investigated GaAs crystals doped with Zn and Cd and found the magnitude of the signals to be very stress sensitive for Zn but less for Cd doped GaAs.

At present a theoretical interpretation of the results still seems to be rather ambiguous as the exact positions of the energy bands and the band–band transition probabilities are not yet sufficiently known. So, more experimental results will have to be waited for.

References

Bemski, G., 1960a, Phys. Rev. Letters **4**, 62.
Bemski, G., 1960b, Phys. Rev. **118**, 1534.
De Wit, M. and T. L. Estle, 1962, Bull. Am. Phys. Soc. **7**, 449.
De Wit, M. and T. L. Estle, 1963, Phys. Rev. **132**, 195.
Duncan, W. and E. E. Schneider, 1963, Phys. Letters **7**, 23.
Fedotov, S. P., V. A. Presnov and V. K. Bazhenov, 1968, Soviet Phys. Solid State **9**, 11.
Kane, E. O., 1957, Phys. Chem. Solids **1**, 249.
Kaplan, D. and M. Guéron, 1965, Compt. Rend. **260**, 2766.
Konopka, J., 1967, Phys. Letters **26** A, 29.
Roth, L. M., B. Lax and S. Zwerdling, 1959, Phys. Rev. **114**, 90.
Title, R. S., 1963, IBM J. Res. Develop. **7**, 68.
Title, R. S., 1967, Phys. Rev. **154**, 668.
Woodbury, H. H. and G. W. Ludwig, 1961, Bull. Am. Phys. Soc. **6**, 118.

7 | GROUP IV ELEMENTS AND COMPOUNDS

The elements silicon and germanium are of great importance in solid state physics and have found numerous applications in modern electronic devices. Therefore these crystals have been subjected to an enormous amount of investigations, both from the aspects of basic research and of technical development.

Spin resonance studies have brought more insight in the behaviour of these crystals, although the principal results in this field have been obtained by electrical and optical measurements. While the latter have shown the decisive influence of small concentrations of impurities, spin resonance techniques have proved extremely useful for analysing the detailed structure of these defects.

Silicon and germanium can be prepared as big single crystals with high purity and can be doped with desired impurities. The lattice atoms have no magnetic moments except the $I = \frac{1}{2}$ nucleus of the 4.7% abundant Si^{29} isotope. So ESR lines are quite narrow; the line widths are of the order of 1 G as compared to often some hundreds G for F centres in alkali halides (see chapter 4). A further characteristic feature for the spin resonance results is the tetrahedral symmetry about each lattice site in the diamond structure of these crystals.

The spin resonance work up to 1961 is covered in the review article by Ludwig and Woodbury (1962). This article can serve as an excellent introduction to this field although of course some newer results are lacking. The book by Corbett (1966) gives a quite up to date account of the results on radiation induced defects.

7.1. Silicon

Silicon has been most extensively studied by spin resonance as it has been by other methods too. The knowledge obtained from other work can be used for better interpretation of the ESR results and vice versa. Our treatment here will concentrate on shallow impurity states, followed by short sections on other defects.

7.1.1. SHALLOW DONORS

The group V elements P, As, Sb and Bi have one more valence electron than Si. When incorporated into Si crystals, this electron has a rather low binding energy (< 0.1 eV). Therefore, these impurities are easily ionized and cause the n type behaviour of donor doped Si crystals. While these ionized impurities are diamagnetic, the neutral ones, which are only stable at low temperature, are paramagnetic and their ESR can easily be detected.

For example fig. 7.1 shows the ESR of a P doped Si crystal at 20 K. Two hyperfine lines ($I = \frac{1}{2}$ for P^{31}) are clearly resolved. The lines are inhomogeneously broadened due to interaction with the 4.7% abundant Si^{29} nuclei. The line width in crystals enriched to 99.9% with the nonmagnetic Si^{28} isotope reduces to 0.24 G from the normal 2.5 G. At liquid helium temperatures, the relaxation times are rather long so that detection of the ESR dispersion signal is preferred. The spin Hamiltonian for these shallow donors is rather simple:

$$H = g\mu_B \, S \cdot H + A \, S \cdot I.$$

The g and A tensors are nearly isotropic. While the g factors are very close to 2.0, the different donors can easily be distinguished by the number and spacing of their hyperfine lines (see table 7.1).

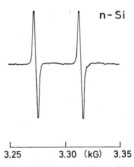

Fig. 7.1. ESR of P doped silicon ($T = 20$ K).

TABLE 7.1

Shallow donors in silicon. Data taken from Feher (1959)

Donor	I	A $(10^{-4}\ cm^{-1})$	Line width (G)
P^{31}	$^1/_2$	39.2	2.5
As^{75}	$^3/_2$	66.2	2.9
Sb^{121}	$^5/_2$	62.3	2.3
Sb^{123}	$^7/_2$	33.9	2.3
Bi^{209}	$^9/_2$	492.1	4.5

As the A values give $|\psi(0)|^2$, a comparison with theoretial calculations of the donor electron wave function is of interest. Luttinger and Kohn (1955) suggested a wave function of the form

$$\psi(r) = \sum_{i=1}^{6} \alpha^i\, F^i(r)\, \psi(k^i, r),$$

where for the ground state the $F^i(r)$ are 1s hydrogen functions and the $\psi(k^i, r)$ are Bloch functions at the ith minimum of the Si conduction band. With some additional assumptions, the calculated $|\psi(0)|^2$ value of P can be brought near to the measured one.

Further information about the spread of the donor wave function in the lattice is obtained by the ENDOR technique (see chapter 2). In fact, it was with P doped silicon crystals that Feher (1959) introduced this new technique into spin resonance work. Fig. 7.2 gives some of his registrations. The ENDOR transitions for the P^{31} nuclei can be seen at 53 and 65 MHz (the difference of 12 MHz being the double nuclear resonance frequency

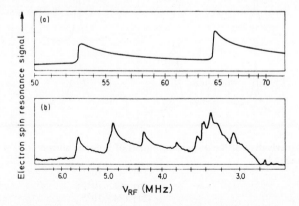

Fig. 7.2. ENDOR of P doped silicon ($T = 1.25$ K). (After Feher, 1959.)

of P^{31} in the applied field of 3000 G). The lines between 3 and 6 MHz are the weaker transitions from the 4.7% abundant Si^{29} nuclei in the neighbourhood of the P atoms. These interactions cause the width of the normal ESR lines. From these data, Feher was able to map the donor wave function in the lattice. It turns out that its amplitude is smaller at the nearest neighbour positions than in some distance. Comparison with theory leads to the estimation that the conduction band minimum in Si is located at $k_0 = 0.85\ k_{max}$.

Very interesting results were obtained in higher doped samples. Maekawa and Kinoshita (1965) investigated P doped Si with concentrations from 3×10^{16} to 3×10^{19} cm^{-3}. Fig. 7.3 gives some of their results. In the lower concentration range, exchange coupled lines of clusters of two and three P atoms can be seen between the two hyperfine lines of isolated P atoms. With higher concentrations, these two hyperfine lines vanish and

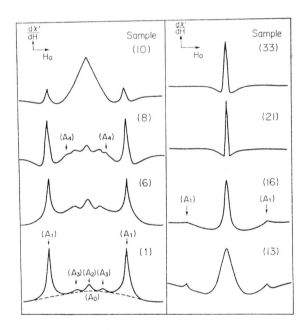

Fig. 7.3. ESR (derivative of the dispersion signal) of P doped silicon (T = 4.2 K). P concentration (in cm^{-3}):

(1) 4×10^{16};	(13) 47×10^{16};
(6) 12×10^{16};	(16) 63×10^{16};
(8) 25×10^{16};	(21) 174×10^{16};
(10) 32×10^{16};	(33) 396×10^{16}.

(After Maekawa and Kinoshita, 1965.)

a sharp line (width ≈ 1 G) arises. This line is ascribed to localized electrons with quick hopping motions. In the high concentration range, with metallic conduction behaviour of the crystals, the line broadens again. It could arise from the nonlocalized electrons in the impurity band (see chapter 5).

The g factor of the donor ground state should be isotropic as an average

$$g = \tfrac{1}{3} g_{\parallel} + \tfrac{2}{3} g_{\perp}$$

over the contributions of the six conduction band valleys. In crystals subjected to uniaxial stress, these contributions are different and $g_{\parallel}$ and $g_{\perp}$ can be determined, $g_{\parallel} - g_{\perp}$ being of the order of 10^{-3} (Wilson and Feher, 1961).

The donor resonances in Si give excellent possibilities to study ESR relaxation and saturation behaviour. Because at low temperatures the relaxation times are long (eventually more than 1000 s!), the populations of the spin levels can be inverted and maser action is observed. Nuclear polarization in these crystals has been extensively studied (see Abragam, 1961).

7.1.2. SHALLOW ACCEPTORS

In contrast to donors, acceptor resonances could only be detected in crystals subjected to uniaxial stress (Feher et al., 1960). It is believed that this is caused by the degeneracy of the valence bands in Si at $k = 0$, which can be lifted by stress. Fig. 7.4 shows the appearance of a single ESR line in B doped Si when stress is applied at low temperature. Ludwig and Woodbury (1961) have studied the spin–lattice relaxation of B in Si as a function of stress. The relaxation time becomes very short as stress is reduced, so that the resulting line broadening prevents the detection of the ESR lines at zero stress.

7.1.3. TRANSITION METAL IONS

These defects are rather unimportant for the technical applications of Si crystals but give fine ESR spectra (Woodbury and Ludwig, 1960). It is interesting to note that manganese has been studied in four different states of charge: Mn^{2+}, Mn^+, Mn and Mn^-! For details, the reader is referred to Ludwig and Woodbury (1962).

Several impurity pairs formed by transition metal ions and acceptors have also been studied by resonance methods.

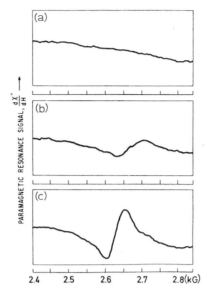

Fig. 7.4. ESR of B acceptors in silicon after application of uniaxial stresses (a) 0 kg/cm^2, (b) 300 kg/cm^2, (c) 900 kg/cm^2 ($T = 1.3$ K, $\nu = 9.065$ GHz). (After Feher et al., 1960.)

7.1.4. RADIATION DAMAGE

By bombardment with fast particles (preferably electrons), many different defects can be produced in Si crystals and have been studied by resonance methods. The book by Corbett (1966) gives a rather complete survey.

Only one example, the Si-A centre (Si-B1 in new nomenclature) is mentioned here. It is generated in n-Si by electron bombardment at room temperature. Its concentration was found to be lower in crystals prepared by the floating zone technique than in those pulled from quartz crucibles. From this, one can conclude that oxygen is involved with this centre. A model is given in fig. 7.5: An interstitial oxygen atom is associated with a lattice vacancy formed by the electron irradiation. Two neighbouring Si atoms are bonded to the oxygen, the two other Si atoms to each other. The latter pair of Si atoms traps an electron in an antibonding state and acts as an acceptor. The ESR spectrum consists of several lines resulting from centres with Si^{28}–Si^{28}, Si^{28}–Si^{29} and Si^{29}–Si^{29} pairs in due intensities. As the two latter pairs give hyperfine interactions, it can be calculated that 70% of the unpaired electron wave function occupies the antibonding state, while the rest is spread over the neighbourhood. This model can further account

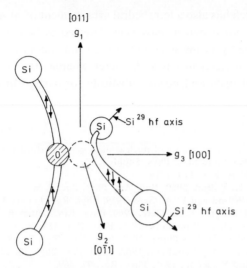

Fig. 7.5. Model of the Si-A centre. (After Watkins and Corbett, 1961.)

for stress experiments. A polarization of the centres and of the infrared oxygen absorption is found.

7.2. Germanium, diamond and SiC

Only few defects have been detected in germanium by resonance methods. The reason for this is probably the large spin–orbit interaction, which leads to short relaxation times. Feher has found the ESR of P and As donors (Feher, 1959; Feher et al., 1957, 1960; Feher and Wilson, 1960). The spectra show hyperfine structure, which is averaged out at higher concentrations. From stress experiments, the g tensor for a single valley is calculated to be

$$g_{\parallel} = 0.87, \qquad g_{\perp} = 1.92,$$

which corresponds well to a theoretical estimate by Roth and Lax (1959).

Spin resonance of Ni^- in Ge has been reported by Ludwig and Woodbury (1957), that of Mn^{2-} by Watkins (1957). In diamond, the spin resonance of a nitrogen donor centre has been found by Smith et al. (1959). The spectra show the three line hyperfine structure due to interaction with the N^{14} nucleus ($I = 1$). The symmetry axis of this interaction may be each of the four nearest neighbour directions. Therefore the nitrogen is concluded to occupy a substitutional position.

In SiC, which has also a tetrahedral arrangement of atoms, ESR spectra of nitrogen and boron defects have been detected by Woodbury and Ludwig (1961). From the hyperfine structure of the resonances the authors conclude that the nitrogen electron has s character whereas the boron electron has p character. Nitrogen and boron substitute for carbon in the SiC lattice.

References

Abragam, A., 1961, The principles of nuclear magnetism (Oxford Univ. Press, London).
Corbett, J. W., 1966, Solid State Phys. Suppl. 7.
Feher, G., 1959, Phys. Rev. **114**, 1219.
Feher, G. and D. K. Wilson, 1960, Bull. Am. Phys. Soc. **5**, 60.
Feher, G., D. K. Wilson and E. A. Gere, 1959, Phys. Rev. Letters **3**, 25.
Feher, G., J. C. Hensel and E. A. Gere, 1960, Phys. Rev. Letters **5**, 309.
Ludwig, G. W. and H. H. Woodbury, 1959, Phys. Rev. **113**, 1014.
Ludwig, G. W. and H. H. Woodbury, 1961, Bull. Am. Phys. Soc. **6**, 118.
Ludwig, G. W. and H. H. Woodbury, 1962, Solid State Phys. **13**, 223.
Luttinger, J. M. and W. Kohn, 1955, Phys. Rev. **97**, 869.
Maekawa, S. and N. Kinoshita, 1965, J. Phys. Soc. Japan **20**, 1447.
Roth, L. M. and B. Lax, 1959, Phys. Rev. Letters **3**, 217.
Smith, W. V., P. P. Sorokin, I. L. Gelles and G. J. Lasher, 1959, Phys. Rev. **115**, 1546.
Watkins, G. D., 1957, Bull. Am. Phys. Soc. **2**, 345.
Watkins, G. D. and J. W. Corbett, 1961, Phys. Rev. **121**, 1001.
Wilson, D. K. and G. Feher, 1961, Phys. Rev. **124**, 1068.
Woodbury, H. H. and G. W. Ludwig, 1960, Phys. Rev. **117**, 102.
Woodbury, H. H. and G. W. Ludwig, 1961, Phys. Rev. **124**, 1083.

8 | OTHER CRYSTALS

8.1. Centres in simple oxides

In the following, compounds such as MgO, CaO, SrO, BaO and Al_2O_3 are considered as simple oxides, in contrast to the more complicated oxides, having the structure of spinels, garnets or perovskites. Results on beryllium oxide and zinc oxide have been collected together with the data of the II–VI compounds. These substances may be regarded as semiconductors with large band gap rather than as insulators.

The resonance data obtained from transition metal ions in the crystals are not discussed in detail but as far as possible are summarized in tables. The most important facts are mentioned and may serve as a comparison for the data found for the multiple oxides.

In some respects, the crystals of the simple oxides and sulphides of face centred cubic structure are ideal hosts for impurity centres and offer a simpler system for the ESR study of many centres than the traditional alkali halides. Of the lattice nuclei, only a few isotopes with little abundance possess a nuclear nonzero spin, and therefore simple hyperfine structure patterns and small linewidths can be expected for the ESR spectra. In the alkali halides, ENDOR techniques were required to resolve many of the inhomogeneously broadened lines of the colour centres, and the spectra indicated interactions with several successive shells of paramagnetic alkali halide ions.

(i) It is easy to incorporate transition elements in all oxides. Rare earth elements substitute both for the cations and anions in CaO and SrO, but only to a rather small amount in MgO and BaO.

(ii) All iron group elements can be substituted for the aluminium ion in

Al_2O_3. The spin Hamiltonian is characterized by a distorted octahedron with a weak trigonal field superimposed.

(iii) Unusual valence states can be stabilized in the hosts, e.g., Fe^+, Ni^+, Co^+. Charge compensation does not necessarily occur in the neighbourhood.

(iv) The crystal field environment may be of octahedral, tetragonal or cubic symmetry, the symmetry showing up in the angular dependence of the spectra.

(v) Most centres are photosensitive to irradiation. Conversion from one charge state to another can be easily induced.

ESR data for alkaline earth oxides and for Al_2O_3 are summarized in tables 8.1–8.4.

TABLE 8.1

ESR data for transition elements in alkaline earth oxides

Ion	Host	T (K)	g	A $(10^{-4}$ $cm^{-1})$	a $(10^{-4}$ $cm^{-1})$	References
V^{2+}	MgO	290	1.9800	−75.1	–	Van Wieringen and Rensen (1963)
	CaO	20	1.9683	76.2	–	Low and Rubins (1962b)
Cr^{3+}	MgO	290	1.9782	16.0	798	Henderson and Hall (1967)
	CaO	77	1.9732	17.0	–	Low and Rubins (1962b)
	SrO	4.2	1.9686	17.3	–	Low and Suss (1964b)
Mn^{2+}	MgO	290	2.0014	−81.0	18.66	Low (1957)
	CaO	20	2.0011	−81.7	6.0	Low and Rubins (1962b)
	SrO	290	2.0012	−78.7	4.3	Holroyd and Kolopus (1963)
Fe^{3+}	MgO	290	2.0037	10.1	205	Auzins et al. (1962)
	CaO	20	2.0059	10.5	65.2	Low and Rubins (1962a)
	SrO	20	2.0063	11.3	–	Auzins et al. (1962)
Fe^{2+}	MgO	4.2	3.4277	–	–	Shuskus (1964)
	CaO	4.2	3.298	–	–	Shuskus (1964)
Fe^+	MgO	20	4.15	–	–	Auzins et al. (1962)
	CaO	4.2	4.1579	33.9	–	Low and Suss (1964a)
Co^{2+}	MgO	4.2	4.2785	97.8	–	Low (1958)
	CaO	4.2	4.3747	131.5	–	Low and Suss (1964a)
Co^+	MgO	77	2.1728	54.0	–	Low and Suss (1965)
Ni^{3+}	SrO	4.2	4.36	–	–	Low and Suss (1964b)
Ni^{2+}	MgO	4.2	2.2145	8.3	–	Orton et al. (1960)
	CaO	20	2.327	–	–	Low and Rubins (1962a)
Ni^+	MgO	77	2.1693	–	–	Auzins et al. (1962)
	CaO	77	2.2814	–	–	Low and Suss (1964b)
Cu^{2+}	MgO	77	2.190	19	–	Auzins et al. (1962)
	CaO	4.2	2.2223	29.1	–	Low and Suss (1963)
Ru^-	MgO	77	2.1697	–	–	Auzins et al. (1962)
Rh^0	MgO	77	2.1708	–	–	Auzins et al. (1962)
Pd^+	MgO	77	2.1698	–	–	Auzins et al. (1962)

TABLE 8.2

ESR parameters for rare earth ions in alkaline earth oxides

Ion	Crystal	T (K)	g	A $(10^{-4}$ cm$^{-1})$	b_4^0 $(10^{-4}$ cm$^{-1})$	b_6^0 $(10^{-4}$ cm$^{-1})$	References
Eu^{2+} (f^7)	CaO	4.2	1.9918	$A^{151} = 30.16$ $A^{153} = 13.46$	25.7	-15.5	Shuskus (1962)
	SrO	77	1.991	$A^{151} = 29.9$ $A^{153} = 13.2$	< 0.5	–	Calhoun and Overmeyer (1962)
Gd^{3+} (f^7)	CaO	4.2	1.9925	–	12.2	-1.19	Shuskus (1962)
	SrO	4.2	1.989	–	5.8	–	Low and Suss (1964b)
Dy^{3+} (f^9)	CaO	20	6.60	–	–	–	Low and Offenbacher (1965)
Er^{3+} (f^{11})	MgO	20	4.03	–	–	–	Low and Offenbacher (1965)
	CaO	4.2	$g_{\parallel} = 4.73$ $g_{\perp} = 7.86$	–	–	–	Low and Rubins (1963)
Yb^{3+} (f^{13})	CaO	20	2.585	$A^{171} = 698$	–	–	Low and Rubins (1963)
	SrO	4.2	2.578	–	–	–	Low and Offenbacher (1965)

TABLE 8.3

ESR parameters for transition group ions in Al$_2$O$_3$

Ion	T (K)	g	$g_{\parallel}$	$g_{\perp}$	A $(10^{-4}$ cm$^{-1})$	B $(10^{-4}$ cm$^{-1})$	b_2^0 $(10^{-4}$ cm$^{-1})$	References
Ti^{3+}	4.2	–	1.067	≤ 0.1	–	–	–	Kornienko and Prokhorov (1960)
V^{2+}	300	–	1.991	1.991	-73.5	-74.3	-1601.2	Laurance and Lambe (1963)
V^{4+}	300	–	1.97	1.97	1.32	–	–	Lambe and Kikuchi (1960)
Cr^{3+}	300	–	1.984	1.984	16.2	16.2	-1908	Geusic (1956)
Mn^{2+}	300	–	2.0017	2.003	79.6	78.8	194.2	Kornienko and Prokhorov (1958)
Mn^{4+}	4.2	–	1.9937	1.9936	70.0	70.0	1957	Geschwind et al. (1962)
Fe^{3+}	4.2	2.003	–	–	–	–	1719	Bogle and Symmons (1959)
Co^{2+}	4.2	–	2.292	4.947	32.4	97.2	–	Zverev and Prokhorov (1961)
Ni^{2+}	290	–	2.1957	2.1859	–	–	13760	Marshall and Reinberg (1960)
Ni^{3+}	4.2	2.146	–	–	–	–	13287	Geschwind and Remaika (1962)
Cu^{3+}	1.4	–	2.0784	2.0772	Cu$^{63} = 64.3$ Cu$^{65} = 68.9$	60.0 64.3	-1883.8	Blumberg et al. (1963)
Gd^{3+}	290	1.9912	–	–	–	–	1032.9	Geschwind and Remaika (1961)
Ru^{3+}	20	–	2.006	2.430	–	–	–	Geschwind and Remaika (1962)

TABLE 8.4

ESR parameters of F centres in alkaline earth oxides

Host	g	A $(10^{-4} \text{ cm}^{-1})$	References
MgO	2.0023	3.72	Wertz et al. (1957)
CaO	2.0001	–	Auzins et al. (1963)
SrO	1.9845	13.55	Culvahouse et al. (1965)
BaO	1.936	59.7	Carson et al. (1959)

From the great number of alkaline earth sulphides and selenides, practically no data have up to now been reported though spectra of F centres in powder samples have been observed. Evidently, the problem is still the preparation of reasonable good single crystals, a problem which we also find with other substances.

A great number of colour centres and related defects can be produced in the alkaline earth oxides by neutron bombardment or by additive colouring (Wertz et al., 1957, 1961, 1962; Carson et al., 1959). The spectrum of the F centre in MgO is a very clear example for F centre resonances. One obtains a hyperfine structure of six lines coming from those F centres which have one Mg^{25} nucleus ($I = \frac{5}{2}$) in the first shell of neighbours, superimposed on a single line caused by interaction with the nuclei having nuclear spin $I = 0$. HFS with Mg^{25} nuclei in the next shells is also resolved. Fig. 8.1 shows an ESR spectrum for the F centre in MgO, together with the HFS contributions from the various combinations of neighbouring Mg^{25} nuclei indicated below the curve. F centres in BaO (Carson et al., 1959) and in SrO (Culvahouse et al., 1965) behave similarly, except that the HFS interaction is with Ba and Sr isotopes.

In most of the substances, additional ESR spectra are found which are believed to be caused by centres which have an electron trapped at a vacancy pair (Auzins et al., 1963). In CaO, a centre analogous to the M centre triplet state in alkali halides, the F_t centre, has been detected by ESR (Tanimoto et al., 1965). Further numerous aggregate and hole centres have been found to occur in these crystals after irradiation (Wertz et al., 1961, 1962; Wertz and Auzins, 1965; Auzins et al., 1963; Kirklin et al., 1965). Some of the above centres have also been analysed by ENDOR. An up to date survey of the defects caused by irradiation with γ-rays on crystals of the alkaline earth oxides has been given by Low and Suss (1966). In addition, many transition metal ions have been identified by ESR.

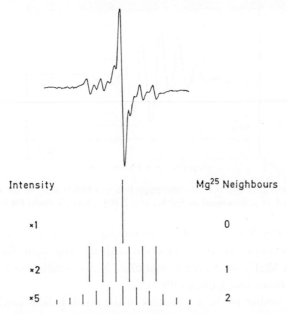

Fig. 8.1. The ESR spectrum of F centres in MgO; the field H along a $\langle 111 \rangle$ axis. The central line belongs to F centres that are surrounded only by Mg^{24} and Mg^{26} nuclei ($I = 0$). Besides this, one recognizes six equidistant weaker lines of the F centre with one Mg^{25} neighbour and several of the 11 still weaker lines of F centres with two Mg^{25} neighbours. (After Wertz et al., 1957.)

8.2. Alkaline earth fluorides

During the last time, numerous results have been reported on transition group ions deliberately added as impurities to alkaline earth fluorides. Most data are available for CaF_2; however, also MgF_2, BaF_2, SrF_2 and even CdF_2 have become compounds of interest to ESR. As an increasing number of results is to be expected, a detailed discussion of the particular results still seems to be premature. The spectra can all be understood with the usual spin Hamiltonian formalism applied to the CaF_2 lattice. F centres and a number of hydrogen centres in the alkaline earth fluorides have been more extensively investigated.

Additively coloured MgF_2, CaF_2 and BaF_2 crystals or those which have been irradiated by X-rays show the ESR of colour centres (Arends, 1965; Blunt and Cohen, 1967). The spectrum in CaF_2 reveals a well resolved hyperfine structure with six next nearest fluorine nuclei (see fig. 8.2). In BaF_2, additional splitting occurs due to the Ba^{135} and Ba^{137} ($I = \frac{3}{2}$) isotopes in

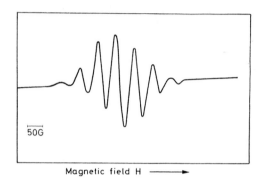

Fig. 8.2. The derivative of the ESR absorption line as a function of H for colour centres in CaF_2, measured at 300 K; $H \parallel [110]$. (After Arends, 1965).

the first shell (fig. 8.3). ENDOR measurements on F centres in CaF_2 (Hayes and Stott, 1967) exhibit the interaction with the first shell fluorine nuclei. M centres in MgF_2 have been found after X-ray irradiation at room temperature by Blunt and Cohen (1967).

Centres similar to the U_2 centres in the alkali halides can be produced in CaF_2, SrF_2 and BaF_2 crystals which have been heated in an atmosphere of hydrogen followed by X-ray irradiation (Hall and Schumacher, 1962; Welber, 1964). One observes a large doublet splitting caused by the interaction of the unpaired electron with the hydrogen proton. Each doublet line is further split into nine lines due to interaction with the eight equivalent nearest neighbour fluorine nuclei (fig. 8.4). ENDOR measurements give the interactions with further shells. X-ray irradiation of hydrogen doped CaF_2 crystals at low temperature leads to substitutional hydrogen atoms on fluorine sites (Bessent et al., 1967). A hole centre similar to the V_k centre in alkali

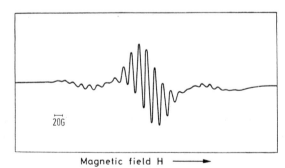

Fig. 8.3. The derivative of the ESR absorption line as a function of H for colour centres in BaF_2, measured at 80 K; $H \parallel [110]$. (After Arends, 1965.)

TABLE 8.5

ESR parameters of irradiation induced centres in the alkaline earth fluorides

Centre	Host	g	Nucleus	a/h (MHz)	b/h (MHz)	T (K)	References
Substitutional hydrogen atom	CaF_2	2.0023	H	1439.5	–	77	Bessent et al. (1967)
			F_I	−91.2	34.8		
			F_{II}	0	1.2		
			F_{III}	1.3	1.3		
			F_{IV}	0.8	0.8		
Interstitial hydrogen atom	CaF_2	2.0029	H	1460.3	–	300	Hall and Schumacher (1962)
			F_I	103.9	34.9		
			F_{II}	0.4	0.9		
			F_{III}	0	0.4		
	SrF_2	2.0053	H	1443	–	300	Welber (1964)
			F_I	75	29		
	BaF_2	2.0023	H	1439.5	–	300	Welber (1964)

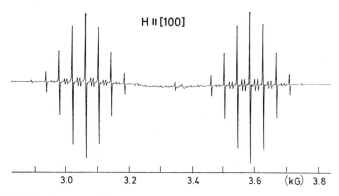

Fig. 8.4. The ESR spectrum of H atoms on interstitial sites in CaF_2, H along a [100] axis. The two groups of nine equidistant lines belong to the two orientations of the proton spin. The doublets between the strong lines are forbidden transitions; the sharp line near 3.35 kG is a g mark. (After Hall and Schumacher, 1962.)

halides has been investigated by ESR and ENDOR (Twidell and Hayes, 1962; Marzke and Mieher, 1965; Kazumata, 1969). ESR parameters of the irradiation induced hydrogen centres in the alkaline earth fluorides are compiled in table 8.5.

8.3. Complex oxides

A great number of complex oxides, such as rutile (TiO_2) barium titanate ($BaTiO_3$), spinel ($MgAl_2O_4$) and garnet ($Y_3Fe_5O_{12}$), are of relative importance for different fields of physics. An interesting feature common to all the different substances of these types is their either electric or magnetic behaviour as far as account is taken of the cooperative properties of the compounds. These phenomena are observed if the complex oxides are in their concentrated magnetic form. The tantalates and aluminates are ferro-electric, the importance of the spinels and garnets is mainly based on their magnetic properties. Many of the compounds show phase transitions which may affect the cooperative phenomena.

In general, the bulk magnetic behaviour is difficult to calculate, whereas the properties of isolated magnetic ions are much better known. To some extent, spin resonance of single defects in these crystals has contributed to the understanding of the behaviour of concentrated magnetic complexes. In some cases it permitted the determination of the point symmetry of particular magnetic ions, in others the number of inequivalent magnetic sites or distortions from simple symmetry. Often spin resonance has also provided better understanding of the degree of order in this group of compounds, and the axes of the individual substructures have been determined from the parameters of the spin Hamiltonian. Progress has also been made in calculating the 'single ion' contribution to the overall anisotropy energy.

In the following, we briefly summarize the main results obtained from spin resonance. We do not attempt to be exhaustive in these particular fields and for details the reader is referred to the excellent review article of Low and Offenbacher (1965).

8.3.1. RUTILE (TiO_2)

We shall be interested in rutile as it seems that the ferroelectric behaviour of some of the titanates is linked with the properties of titanium dioxide. Here the Ti^{4+} ion is surrounded by six octahedrally coordinated oxygen ions. The crystal structure of rutile is based on a tetragonal unit cell of which the TiO_3 octahedron is a subunit. The substance has a large dielectric constant and a relatively low loss factor in the microwave and infrared regions. Experimentally, this fact allows that the crystal may act as its own cavity. The pure TiO_2 single crystal, normally grown by means of the flame fusion method, is an insulator. Nonstoichiometric crystals may be semiconducting with a variable energy gap depending on stoichiometry. In the unit cell, there are two octahedral sites which differ by a rotation of 90° around the tetragonal axis of the crystal. They give rise to two ESR spectra for a general orientation of the crystal axis as referred to the magnetic field. Impurity ions may be incorporated either substitutionally or as interstitials at four identical but not equivalent sites. Most of the ions with d^n configuration as well as many rare earth ions can be incorporated despite the fact that the ionic radius may be much larger than the space available. More than in other ionic lattices, unusual valence states are found although charge compensation does not necessarily take place in the neighbourhood of the impurity centre.

The ESR spectra can be explained with the usual spin Hamiltonian. Effects of the large polarizability of rutile on the spectra have not been observed. For some impurities, superhyperfine interaction with the Ti^{47} and Ti^{49} ($I = \frac{5}{2}$ and $\frac{7}{2}$) nuclei is observed. ESR data for transition elements in rutile are collected in table 8.6.

Recently, data on other rather simple oxides such as VO_2, ZrO_2, SnO_2 or GeO_2 are reported in an increasing number, which may be due to success in better sample preparation.

TABLE 8.6

ESR parameters for transition group ions in titanium dioxide

Ion	T (K)	g_x	g_y	g_z	A_x (10^{-4} cm^{-1})	A_y (10^{-4} cm^{-1})	A_z (10^{-4} cm^{-1})	b_2^0 (10^{-4} cm^{-1})	b_2^2 (10^{-4} cm^{-1})	References
Ti^{3+}	4.2	1.974	1.977	1.941	–	–	43	–	–	Chester (1961b)
V^{4+}	4.2	1.915	1.956	1.912	31	142	16.7	–	–	Gerritsen et al. (1960)
Cr^{3+}	4.2	1.97	1.97	1.97	–	–	16.7	6800	6800	Gerritsen et al. (1960)
Mn^{4+}	4.2	1.991	1.995	1.990	72.3	70.3	72.7	4000	3900	Yamaka and Barnes (1964)
Mn^{3+}	.77	2.00	2.00	1.99	84.5	52.8	80.6	3400	3480	Gerritsen and Sabisky (1963)
Fe^{3+}	1.4	2.000	2.000	2.000	–	–	–	6780	2210	Carter and Okaya (1960)
Co^{2+}	4.2	5.88	2.19	3.75	142.8	39.1	25.0	–	–	Yamaka and Barnes (1962)
Ni^{3+}	4.2	2.084	2.254	2.085	–	–	–	–8300	1370	Gerritsen and Sabisky (1962)
Ni^{2+}	4.2	2.10	2.20	2.10	–	–	–	–8300	1370	Gerritsen and Sabisky (1962)
Cu^{2+}	77	2.105	2.344	2.093	–19	–88	–29	–	–	Gerritsen and Sabisky (1962)
Gd^{3+}	295	$g = 1.994$			–	–	–	120.9	22.8	Yamaka (1963)
Er^{3+}	4.2	$g_\perp < 0.1$		15.1	495	–	–	–	–	Gerritsen and Sabisky (1963)
Ce^{3+}	4.2	4.39	2.06	3.86	–	–	–	–	–	Chester (1961a)
Nb^{4+}	25	1.973	1.981	1.948	8.04	1.75	2.1	–	–	Chester (1961a)
Mo^{5+}	77	1.812	1.913	1.788	2.47	65.8	30.5	–	–	Ru-Tao Kui (1962)
W^{5+}	63	1.595	1.473	1.446	92.0	40.5	63.9	–	–	Chang (1964)

8.3.2. PEROVSKITES

Compounds of the composition ABO_3 (barium titanate $BaTiO_3$ is a well known example) are named perovskites and have interesting electric properties. $BaTiO_3$ and $PbTiO_3$ are ferroelectric at room temperature. $SrTiO_3$ is paraelectric over nearly the whole temperature range. In reduced form, strontium titanate is a semiconductor which becomes superconducting at very low temperatures. $LaMnO_3$ is an antiferromagnetic substance.

In the perovskite structure, the ion B^{4+} is surrounded by an octahedron of oxygen ions whose axes are along the cubic axes. This subunit is similar to the TiO_6 complex in rutile. The A ion has twelve nearest oxygen and eight B neighbours. A great number of the perovskite type compounds show phase transitions at well defined temperatures. The nature of these transitions and the change in domain structures are reflected in the changes of the parameters in the spin Hamiltonian and can partly be studied by incorporation of small fractions of paramagnetic impurities for the positive ions. Up to now, the main reason for the still rather limited number of ESR results is the great difficulty in preparing good single crystals. ESR data for perovskites are collected in table 8.7.

The most thoroughly investigated compound is $BaTiO_3$. The substance shows four phase transitions at definite temperatures together with discontinuous changes in the dielectric constant at these temperatures as well as discontinuities of the lattice constants and thermal expansion. Due to the large dielectric constant, a great number of so called 'forbidden transitions' can be seen. In particular the spectra of trivalent iron and gadolinium have been studied. The spectra show abrupt changes in the spectrum near the Curie temperature at 120 °C. Below the transition temperature, a tetragonal spectrum is observed, the tetragonal splitting being temperature dependent. This is shown in fig. 8.5. An intrinsic centre which can be regarded as an F centre is observed in reduced $BaTiO_3$ crystals. The resonance shows HFS with Ti^{47} and Ti^{49} nuclei (Takeda and Watanabe, 1967).

Strontium titanate has, at room temperature, the ideal perovskite structure. A phase transition is found at 110 K. In contrast to $BaTiO_3$, there is a gradual transition from the axial to a cubic spectrum near the transition temperature (fig. 8.6). Many iron group and rare earth elements can be substituted at the Ti and Sr sites permitting ESR investigations of the octahedral and dodecahedral sites. Nuclear magnetic resonance has revealed a change in the Sr resonance at the phase transition which indicates a displacement of the ion (Weber and Allen, 1963).

Few results have been reported on other crystals having perovskite

TABLE 8.7

ESR parameters for perovskite type materials

Ion	Host	T (K)	g	$g_{\parallel}$	$g_{\perp}$	A (10^{-4} cm^{-1})	b_2^4 (10^{-4} cm^{-1})	b_1^4 (10^{-4} cm^{-1})	References
V^{2+}	$KMgF_3$	77	1.972	–	–	86.2	–	–	Hall et al. (1963)
Cr^{3+}	$KMgF_3$	77	1.9733	–	–	–	–	–	Hall et al. (1963)
	$LaAlO_3$	300	1.9825	–	–	19	450	–	Kiro et al. (1963)
	$SrTiO_3$	300	1.978	–	–	15.8	–	–	Müller (1960)
Cr^+	$KMgF_3$	77	2.000	–	–	23.0	–	2.25	Hall et al. (1963)
Mn^{4+}	$SrTiO_3$	295	1.994	–	–	75	–	–	Müller (1959a)
Mn^{2+}	$KMgF_3$	300	2.0015	–	–	91	–	3.25	Hall et al. (1963)
Fe^{3+}	$BaTiO_3$	425	2.003	–	–	–	–	17	Hornig et al. (1959)
	$SrTiO_3$	1.9	2.004	–	–	–	17.9	115	Müller (1959b)
	$PbTiO_3$	290	–	2.009	5.97	–	–	–	Gainon (1964)
	$KTaO_3$	4.2	–	1.99	6.0	–	–	172.5	Wemple (1963)
	$KMgF_3$	77	2.0031	–	–	–	–	25.8	Hall et al. (1963)
Co^{2+}	$KMgF_3$	4.2	4.28	–	–	–	–	–	Low and Zusman (1963)
Ni^{3+}	$SrTiO_3$	4.2	–	2.110	2.213	–	–	–	Kiro et al. (1963)
Ni^{2+}	$SrTiO_3$	80	2.204	–	–	–	–	–	Kiro et al. (1963)
Ni^+	$SrTiO_3$	20	–	2.029	2.352	–	–	–	Kiro et al. (1963)
Gd^{3+}	$LaAlO_3$	20	1.9909	–	–	–	490	6.64	Low and Zusman (1963)
Gd^{3+}	$BaTiO_3$	300	1.995	–	–	–	293	4.0	Rimai and De Mars (1962b)
	$SrTiO_3$	4.2	1.992	–	–	–	–362	–3.24	Rimai and De Mars (1962b)
Eu^{2+}	$SrTiO_3$	2	1.990	–	–	–	–10.4	106.6	Rimai and De Mars (1962b)

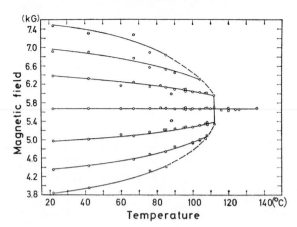

Fig. 8.5. Temperature dependence of the ESR spectrum of Gd^{3+} in barium titanate ($BaTiO_3$). To be noticed is the sudden collapse of the axial spectrum into the cubic spectrum, indicative of a first order phase transition. (After Rimai and De Mars, 1962b.)

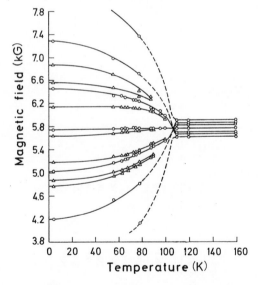

Fig. 8.6. Temperature dependence of the ESR spectrum of Gd^{3+} in strontium titanate ($SrTiO_2$). There is a gradual transition from the axial spectrum to the cubic spectrum near the transition temperature. The phase transition is not of first order. (After Rimai and De Mars, 1962a.)

structure. Lead titanate is ferroelectric at room temperature having a Curie temperature of 490 °C. The dielectric constant varies from a few hundred at room temperature to some thousand at liquid helium temperature.

TABLE 8.8

ESR data for spinels (AB_2O_4)

Ion	Host	T (K)	g	A (10^{-4} cm^{-1})	$b_2{}^0$ (10^{-4} cm^{-1})	$b_4{}^0$ (10^{-4} cm^{-1})	References
Cr^{3+}	$ZnAl_2O_4$	290	—	—	−0.93	—	Brun et al. (1960)
	$MgAl_2O_4$	290	$g_{\parallel} = 1.986$	—	−0.92	—	Brun et al. (1960)
Mn^{2+}	$MgAl_2O_4$	290	2.0015	75.4	—	—	Waldner (1962)
	$ZnAl_2O_4$	290	2.0002	74.9	—	—	Waldner (1962)
	$Li_{0.5}Al_{2.5}O_4$	290	2.0023	77.2	—	—	Kelly et al. (1961)
Fe^{3+}	$Li_{0.5}Al_{2.5}O_4$	290	2.006	—	0.104	0.01	Folen (1960)
	$MgAl_2O_4$	290	2.001	—	0.247	0.02	Brun et al. (1961)

Potassium tantalate ($KTaO_3$) behaves similar to $SrTiO_3$. Data for potassium magnesium fluoride and lanthanum aluminate are also included in table 8.7.

8.3.3. SPINELS

Compounds of the form AB_2O_4, A representing a divalent ion such as magnesium, zinc or cadmium, B representing aluminium or other trivalent ions, are called spinels. The unit cell has a complicated structure containing eight AB_2O_4 molecules and having cubic symmetry. The A sites are of cubic symmetry, the B sites possess an octahedral crystal field. Most crystals of the spinel type have a more or less amount of disorder in the lattice. ESR has contributed considerably to the understanding of the distribution of the magnetic ions over the A and B sites. ESR data for spinels are summarized in table 8.8.

8.3.4. GARNETS

Garnets are compounds of the general formula $A_3B_5O_{12}$, the crystal structure being quite complicated and having overall cubic symmetry. The crystals have low microwave losses and they have been used to study ferri-

TABLE 8.9
ESR parameters for garnets. (Here YGaG stands for $Y_3Ga_8O_{12}$, etc.)

Ion	Host	T (K)	g	$b_2{}^0$ (10^{-4} cm^{-1})	$b_4{}^0$ (10^{-4} cm^{-1})	References
Cr^{3+}	YGaG	290	1.98	3500	–	Carson and White (1961)
	YAlG	300	1.98	2550	–	Geschwind (1961)
Fe^{3+}	YGaG	4.2	2.0047	−880	31	Geschwind (1961)
Pt^{3+}	YAlG	4.2	2.361	–	–	Hodges et al. (1966)

Ion	Host	T (K)	g_x	g_y	g_z	References
Nd^{3+}	YGaG	4.2	2.027	1.251	3.667	Wolf et al. (1962)
	YAlG	4.2	1.733	1.779	3.915	Wolf et al. (1962)
	LuAlG	4.2	1.789	1.237	3.834	Wolf et al. (1962)
Dy^{3+}	LuGaG	4.2	13.45	0.57	3.41	Wolf et al. (1962)
	YGaG	4.2	11.07	1.07	7.85	Wolf et al. (1962)
Er^{3+}	LuGaG	4.2	3.183	3.183	12.62	Wolf et al. (1962)
	YAlG	4.2	7.75	3.71	7.35	Wolf et al. (1962)
Yb^3	YAlG	20	3.87	3.78	2.47	Wolf et al. (1962)
Gd^{3+}	YGaG	300	1.991	–	–	Rimai and De Mars (1962a)
	YAlG	300	1.990	–	–	Rimai and De Mars (1962a)

magnetism. The compound yttrium iron garnet (YIG) has been used for microwave devices. The spontaneous magnetization in these crystals is altered when rare earth ions are substituted for yttrium. Hence, most of the measurements were done with crystals containing rare earth impurity ions. ESR data are collected in table 8.9.

8.4. Alkali hydrides and alkali azides

Mainly F centres and colloids are formed in alkali hydrides and alkali azides by irradiation processes at various temperatures. X-ray irradiation at room temperature produces F centres in NaH. The ESR spectrum shows the hyperfine interaction of the F centre electron with six equivalent sodium nuclei in the first shell. Thus 19 lines with a separation of 26.5 G are observed. The F centres bleach rapidly at about 100 °C. Aside from F centres, colloids are formed in which conduction electron resonance is detected (Doyle and Williams, 1961; Williams, 1962). In LiH, irradiation with electrons, neutrons or short wavelength ultraviolet light at 77 K produces F centres. The resonance spectrum shows an inhomogeneously broadened line with a halfwidth of about 50 G. HFS is not resolved. By warming up, the aggregation process of F centres into colloids can be followed (Lewis and Pretzel, 1961; Pretzel et al., 1962). Doyle (1962) has also studied the ESR of F centres in KH, RbH, CsH and KD. He found a well resolved nearest neighbour hyperfine splitting in all crystals. ENDOR measurements have been performed with these substances too. After irradiation at low temperature, spectra from V type centres appear which have been investigated by Lewis and Pretzel (1961).

Similar effects can be observed in NaN_3 and KN_3 after irradiation with ultraviolet light. Depending on the temperature of irradiation, spectra from F centres with resolved hyperfine structure or from aggregates or from conduction electrons in colloids are obtained (King et al., 1961a,b). After X-ray irradiation at 77 K, spectra of several nitrogen radicals are observed, such as nitrogen atoms and N_2^- molecule ions, the latter forming N_4^- molecule ions after annealing. These centres have all been identified by the number and intensity of their HFS components (Shuskus et al., 1960; Wylie et al., 1962).

8.5. Centres in the ammonium halides

The ammonium halides generally have the CsCl structure with a somewhat complicated unit cell. An F centre is produced in these crystals after irradiation at room temperature with 1 MeV electrons (Patten, 1968). The ESR spectrum shows nine equally spaced lines and the intensity distribution is in good agreement with that expected for an unpaired electron trapped at an anion vacancy and interacting equally with eight hyrogen nuclei.

Irradiation with X-rays, protons or electrons at various temperatures results in the formation of Cl_2 or Br_2 ion defects that are very similar to the V centres in the alkali halides, except that they are oriented along $\langle 100 \rangle$ directions, as opposed to $\langle 110 \rangle$ directions in the alkali halides. These centres were extensively studied by Patten and Marrone (1965). Under the same conditions one observes the spectrum of a hydrazine like defect, i.e. a molecular ion involving two nitrogen and four hydrogen ions (Marquardt and Patten, 1969; Marquardt, 1970).

Many details on irradiation induced centres in the ammonium halides have been reported in a paper by Vannotti et al. (1967).

8.6. Irradiation induced centres in quartz

A number of paramagnetic defects can be produced in quartz by irradiation with neutrons or γ-rays. Partly these centres are associated with impurity ions, e.g. Al, partly they are intrinsic defects. The E_1 and E_2 centres have most extensively been investigated by Weeks (1963) and Castle et al. (1963). As a model for the E_1 centre, an oxygen double vacancy has been suggested with one or three electrons trapped at the Si ion between the vacancies. Hyperfine structure with 4.7% abundant Si^{29} nuclei is resolved. The E_2 centre is supposed to consist of a SiO vacancy, one electron being trapped at the silicon with the broken oxygen bond. Further a proton is trapped in the neighbourhood, of which the hyperfine structure splitting is observed.

8.7. Conduction electron spin resonance (CESR) in metals

Up to now, only little ESR work has been done concerning conduction electrons in metals. Partly because few materials yield detectable spectra but largely because of the success of cyclotron resonance. Both techniques are of interest as they supply information about the electronic states in metals. Line widths, g values and relaxation times allow conclusions about wave functions and scattering processes in metals. However, cyclotron resonance with its information about effective mass and multiple orbits is the most informative.

CESR has been reported in the alkali metals, in beryllium, in copper and in gold, in aluminium and in tin. Usually the short relaxation time results in an undetectable broad signal. With reference to a mechanism proposed by Elliot (1955), Wertz (1955) mentions that one will perhaps not find ESR absorption with heavy metals which have large spin–orbit coupling. Impurities with strong spin–orbit coupling would also interfere with resonance detection. Therefore pure material, perfect crystal lattice and low dislocation concentration are required. Further the skin effect, which reduces the sensitivity of the spectrometer, has to be considered.

The main informations in CESR spectra are the spectroscopic splitting factor and the relaxation time. In metals, the CESR line shape is a function of the sample size. Therefore some theory is required to extract g and T_2 from the asymmetrical lines.

The theory of the effect was worked out by Dyson (1955) and Azbel et al. (1959) and has since been modified by several authors (Gaspari, 1966; Lampe and Platzman, 1966). Briefly, three general experimental situations arise:

(i) Thin case: In small particles or thin films where the thickness is small compared with the skin depth of the microwave, the diffusion time T_D is unimportant and the line has an ordinary symmetrical Lorentz shape.

(ii) Intermediate case: The metal is of about the same thickness as the skin depth, i.e. about 10^{-5} cm. The case is difficult to treat but can be avoided experimentally.

(iii) Thick case: The metal is thick as compared with the skin depth. Feher and Kip (1955) have worked out limiting cases. The line shapes of the absorption and derivative curves in the normal skin effect region are shown in fig. 8.7. The importance of the T_D/T_2 ratio is apparent. Kip and Feher also show how it can be decided from the spectrum that the resonances arise from conduction electrons.

The alkali metals have been studied most extensively. Thin samples have been produced by ultrasonically dispersing the bulk metal in mineral oil or by freezing metal–ammonia solutions. Decomposition of sodium or potassium azides generates metal colloids. Other forms of thin samples have been obtained by evaporation onto dielectric substrate. Thick samples can be prepared by rolling the material under a protective medium.

The best opportunity of detecting CESR signals is offered by investigating thin metal single crystals. CESR of copper, aluminium and tin has been detected in this way by utilizing the 'selective transmission' technique. This uses the fact that in the resonant field, electrons diffusing from one side of a

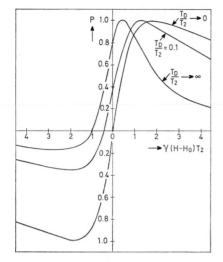
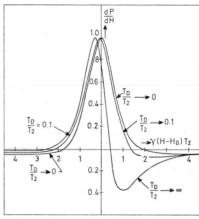

Fig. 8.7. The line shapes of the power absorption P and the derivative of the power absorption dP/dH due to electron spin resonance in thick metal plates for different ratios of diffusion time T_D and relaxation time T_2. (After Feher and Kip, 1955.)

metal (where an rf field is applied) can carry information via their nonequilibrium magnetization to the other side where they radiate power.

Values of g have been determined both by comparison to the free radical DPPH and by simultaneous measurements of the rf frequency and the magnetic field. In table 8.10, the g values, experimental and theoretical g shifts for the pure alkali metals and for other pure metals have been summarized.

TABLE 8.10
The g values of conduction electrons in metals

Metal	g	δg	δg (calculated)	References
Li	2.00229	-2×10^{-6}	$<1 \times 10^{-5}$	Pressley and Berk (1965)
Na	2.0015	-8×10^{-4}	-7×10^{-4}	Ryter (1963)
K	1.9997	-26×10^{-4}	$\approx 20 \times 10^{-4}$	McMillan (1964)
Rb	1.9984	-39×10^{-4}	$\approx 20 \times 10^{-3}$	Walsh et al. (1966)
Cs	2.0055	32×10^{-4}	–	Walsh et al. (1966)
Be	2.0032	9×10^{-4}	–	Orchard-Webb and Cousins (1968)
Mg	2.009	67×10^{-4}	–	Feher and Kip (1955)
Al	1.997	-50×10^{-4}	–	Schultz et al. (1966)
Cu	2.031	29×10^{-3}	–	Schulz and Latham (1965)
Au	2.26	258×10^{-3}	–	Dupree et al. (1967)
Sn	1.9945	-78×10^{-4}	–	Khaikin (1960)

References

Arends, J., 1965, Phys. Stat. Sol. **7**, 805.

Auzins, P., J. W. Orton and J. E. Wertz, 1962, in: Proc. 1st Intern. Conf. on Paramagnetic resonance, Vol. **1** (Academic Press, New York) 90.

Azbel, T., B. Gerasimenko and M. Lifschitz, 1958, Sovjet Phys. JETP English Transl. **35**, 691.

Bessent, R. G., W. Hayes and J. W. Hodby, 1967, Proc. Roy. Soc. London **297**, 1450.

Blumberg, W. E., J. Eisinger and S. Geschwind, 1963, Phys. Rev. **130**, 900.

Blunt, R. F. and M. I. Cohen, 1967, Phys. Rev. **153**, 1031.

Bogle, G. S. and F. H. Symmons, 1959, Proc. Phys. Soc. London **73**, 531.

Brun, E., S. Hofner, H. Loelinger and F. Waldner, 1960, Helv. Phys. Acta **33**, 966.

Brun, E., H. Loelinger and F. Waldner, 1961, Arch. Sci. Geneva **14**, 167.

Calhoun, B. A. and J. Overmeyer, 1962, J. Appl. Phys. **35**, 989.

Carson, J. W. and R. L. White, 1961, J. Appl. Phys. **32**, 1787.

Carson, W., D. F. Holcomb and H. Rüchardt, 1959, Phys. Chem. Solids **12**, 66.

Carter, D. L. and A. Okaya, 1930, Phys. Rev. **118**, 1485.

Castle, J. G., D. W. Feldman, P. G. Klemens and R. A. Weeks, 1963, Phys. Rev. **130**, 577.

Chang, T., 1964, Bull. Am. Phys. Soc. **9**, 568.

Chester, P. F., 1961, J. Appl. Phys. **32**, 866.

Chester, P. F., 1961b, J. Appl. Phys. **32**, Suppl. 2233.

Culvahouse, J. W., L. V. Holroyd and J. L. Kolopus, 1965, Phys. Rev. **140**, A1181.

Doyle, W. T., 1962, Phys. Rev. **126**, 1421.

Doyle, W. T. and W. L. Williams, 1961, Phys. Rev. Letters **6**, 537.

Dupree, R., C. T. Forwood and M. J. A. Smith, 1967, Phys. Stat. Sol. **24**, 525.

Dyson, F. J., 1955, Phys. Rev. **98**, 349.

Elliot, R. J., 1955, Phys. Rev. **96**, 266.

Feher, G. and A. F. Kip, 1955, Phys. Rev. **98**, 337.

Folen, V. J., 1960, J. Appl. Phys. **31**, 166.

Gainon, D. A., 1964, Phys. Rev. A **134**, 1300.

Gaspari, G. D., 1966, Phys. Rev. **151**, 215.

Gerritsen, H. J. and E. S. Sabisky, 1962, Phys. Rev. **125**, 1853.

Gerritsen, H. J. and E. S. Sabisky, 1963, Phys. Rev. **132**, 1507.

Gerritsen, H. J., S. E. Harrison and H. R. Lewis, 1960, J. Appl. Phys. **31**, 1566.

Geschwind, S., 1961, Phys. Rev. **121**, 363.

Geschwind, S. and J. P. Remaika, 1961, Phys. Rev. **122**, 757.

Geschwind, S. and J. P. Remaika, 1962, J. Appl. Phys. **33**, 370.

Geschwind, S., P. Kislink, M. P. Klein, J. P. Remaika and D. L. Wood, 1962, Phys. Rev. **126**, 1684.

Geusic, J. E., 1956, Phys. Rev. **102**, 1252.

Hall, J. L. and R. T. Schumacher, 1962, Phys. Rev. **127**, 1892.

Hall, T. P. P., W. Hayes, R. W. Stevenson and J. Wilkens, 1963, J. Chem. Phys. **38**, 1977.

Hayes, W. and J. P. Stott, 1967, Proc. Roy. Soc. London A **301**, 313.

Henderson, B. and T. P. P. Hall, 1967, Proc. Phys. Soc. London **90**, 511.

Hodges, J. A., R. A. Serway and S. A. Marshall, 1966, Phys. Rev. **151**, 196.

Holroyd, L. V. and J. L. Kolopus, 1963, Phys. Stat. Sol. **3**, 12 K 456.

Hornig, A. W., R. C. Rempel and H. G. Weaver, 1959, Phys. Chem. Solids **10**, 1.

Kazumata, Y., 1969, Phys. Stat. Sol. **34**, 377.

Kelly, R. H., V. J. Folen, M. Mass and W. G. Beard, 1961, Phys. Rev. **124**, 80.

Khaikin, M. S., 1960, Sovjet Phys. JETP English Transl. **39**, 899.

King, G. J., F. F. Carlson, B. S. Miller and R. C. McMillan, 1961a, J. Chem. Phys. **34**, 1499.

King, G. J., F. F. Carlson, B. S. Miller and R. C. McMillan, 1961b, J. Chem. Phys. **35**, 1441.

Kirklin, P. W., P. Auzins and J. E. Wertz, 1965, Phys. Chem. Solids **26**, 1067.

Kiro, D., W. Low and A. Zusman, 1963, in: W. Low, ed., Proc. 1st Intern. Conf. on Paramagnetic resonance, Vol. **1** (Academic Press, New York) 44.

Kornienko, L. S. and A. M. Prokhorov, 1958, Soviet Phys. JETP English Transl. **6**, 620.

Kornienko, L. S. and A. M. Prokhorov, 1960, Soviet Phys. JETP English Transl. **11**, 1189.

Lambe, J. and C. Kikuchi, 1960, Phys. Rev. **118**, 71.

Lampe, M. and P. M. Platzman, 1966, Phys. Rev. **150**, 340.

Laurance, N. and J. Lambe, 1963, Phys. Rev. **132**, 1029.

Lewis, W. B. and F. E. Pretzel, 1961, Phys. Chem. Solids **19**, 139.

Low, W., 1957, Phys. Rev. **105**, 793.

Low, W., 1958, Phys. Rev. **109**, 256.

Low, W. and E. L. Offenbacher, 1965, Solid State Phys. **17**, 200.

Low, W. and R. S. Rubins, 1962a, in: W. Low, ed., Proc. 1st Intern. Conf. on Paramagnetic resonance, Vol. **1** (Academic Press, New York) 79.

Low, W. and R. S. Rubins, 1962b, Phys. Letters **1**, 316.

Low, W. and R. S. Rubins, 1963, Phys. Rev. **131**, 2527.

Low, W. and J. T. Suss, 1963, Phys. Letters **7**, 310.

Low, W. and J. T. Suss, 1964a, Bull. Am. Phys. Soc. **9**, 36.

Low, W. and J. T. Suss, 1964b, Phys. Letters **11**, 115.

Low, W. and J. T. Suss, 1965, Phys. Rev. Letters **15**, 519.

Low, W. and J. T. Suss, 1966, in: Proc. Intern. Conf. on Electron diffraction and the nature of defects (Pergamon, London).

Low, W. and A. Zusman, 1963, Phys. Rev. **130**, 145.

Marquardt, Ch. L., 1970, J. Chem. Phys. **53**, 3248.

Marquardt, Ch. L. and F. W. Patten, 1969, Solid State Commun. **7**, 393.

Marshall, S. A. and A. R. Reinberg, 1960, J. Appl. Phys. **31**, 336.

Marzke, R. and R. L. Mieher, 1965, in: Proc. Intern. Symp. on Color centres, Urbana, Ill.

McMillan, R. C., 1964, J. Phys. Chem. Solids **25**, 773.

Müller, K. A., 1958, Arch. Sci. Geneva **11**, 150.

Müller, K. A., 1959a, Phys. Rev. Letters **2**, 341.

Müller, K. A., 1959b, Helv. Phys. Acta B1, Suppl. 173.

Orchard-Webb, J. H. and J. E. Cousins, 1968, Phys. Letters **28**, 236.

Orton, J. W., P. Auzins and J. E. Wertz, 1960, Phys. Rev. Letters **4**, 128.

Patten, F. W., 1968, Phys. Rev. **175**, 1216.

Patten, F. W. and M. J. Marrone, 1966, Phys. Rev. **142**, 513.

Pressley, R. J. and H. L. Berk, 1965, Phys. Rev. **140**, A 1207.

Pretzel, F. E., E. G. Lewis, E. G. Szklarz and D. T. Vier, 1962, J. Appl. Phys. Suppl. **33**, 510.

Rimai, L. and G. A. de Mars, 1962a, J. Appl. Phys. **33**, 1254.

Rimai, L. and G. A. de Mars, 1962b, Phys. Rev. **127**, 702.

Ru-Tao Kui, 1962, J. Appl. Phys. **128**, 151.

Ryter, Ch., 1963, Phys. Letters **4**, 69.

Schultz, S. and C. Latham, 1965, Phys. Rev. Letters **15**, 148.

Schultz, S., G. Dunifer and C. Latham, 1966, Phys. Letters **23**, 192.

Shuskus, A. J., 1962, Phys. Rev. **127**, 2022.

Shuskus, A. J., 1964, J. Chem. Phys. **40**, 1602.

Shuskus, A. J., C. G. Young, O. R. Gilliam and P. W. Levy, 1960, J. Chem. Phys. **33**, 622.

Takeda, T. and A. Watanabe, 1967, J. Phys. Soc. Japan **23**, 469.

Tanimoto, D. H., W. M. Ziniker and J. O. Kemp, 1965, Phys. Rev. Letters **14**, 645.

Twidell, W. and W. Hayes, 1962, Proc. Roy. Soc. London A **79**, 1295.

Vannotti, L. E., H. R. Zeller, K. Bachmann and W. Känzig, 1967, Phys. Kondens. Materie **6**, 51.

Van Wieringen, J. S. and J. G. Rensen, 1963, in: Proc. 1st Intern. Conf. on Paramagnetic resonance, Vol. **1** (Academic Press, New York) 105.

Waldner, F., 1962, Helv. Phys. Acta **35**, 756.

Walsh Jr., W. M., L. W. Rupp Jr. and P. H. Schmid, 1966, Phys. Rev. Letters **16**, 181.

Weber, M. J. and R. R. Allen, 1963, J. Chem. Phys. **38**, 726.

Weeks, R. A., 1963, Phys. Rev. **130**, 570.

Welber, B., 1964, Phys. Rev. **136**, A 1408.

Wemple, W., 1963, Ph. D. Thesis, Mass. Inst. Technol.

Wertz, J. E., 1955, Chem. Rev. **55**, 829.

Wertz, J. E. and P. Auzins, 1965, Phys. Rev. **139**, A 1645.

Wertz, J. E., P. Auzins, R. A. Weeks and R. H. Silsbee, 1957, Phys. Rev. **107**, 1535.

Wertz, J. E., J. W. Orton, and P. Auzins, 1961, Discussions Faraday Soc. **31**, 140.

Wertz, J. E., J. W. Orton and P. Auzins, 1962, J. Appl. Phys. Suppl. **32**, 322.

Williams, W. L., 1962, Phys. Rev. **125**, 82.

Wolf, W. P., M. Bell, M. T. Hutchins, M. J. M. Leask and A. F. G. Wyatt, 1962, J. Phys. Soc. Japan A, 443.

Wylie, D. W., A. J. Shuskus, C. G. Young, O. R. Gilliam and P. W. Levy, 1962, Phys. Rev. **125**, 451.

Yamaka, E., 1963, J. Phys. Soc. Japan **18**, 1557.

Yamaka, E. and R. G. Barnes, 1962, Phys. Rev. **125**, 1568.

Yamaka, E. and R. G. Barnes, 1964, Phys. Rev. **135**, A 1544.

Zverev, G. M. and A. M. Prokhorov, 1961, Sovjet Phys. JETP English Transl. **12**, 41.

LIST OF SYMBOLS

a	cubic field parameter, isotropic contact term in HFS interaction
A_{ij}	magnetic hyperfine interaction tensor
b	component of the anisotropic hyperfine tensor
B_{ij}	anisotropic hyperfine tensor
B_n^m	coefficient in the crystal field expansion
c	symmetry axis of the crystal field
c	velocity of light
D, E	fine structure coefficients
e	electron charge
E	energy
F	parameter of an orthorhombic crystal field
g	spectroscopic splitting factor
g_I	nuclear g factor
g_s	free electron g factor
h	Planck's constant
H	magnetic field
H_0	resonant field
ΔH	line width at half of the maximum intensity
$\boldsymbol{H}$	Hamilton operator
I	spin of nucleus in units of $h/2\pi$
J	total spin quantum number of an atom
k	Boltzmann's constant
$\boldsymbol{k}$	wave vector
L	electronic orbital angular momentum quantum number of an entire atom or molecule

211

m	nuclear spin quantum number
m_0	rest mass of electron
m^*	effective electron mass
M	magnetisation, quantum number governing the projection of electron spin on the axis of an applied field
N	population number, nuclear number
O_n^m	operator equivalents
P_{ij}	quadrupole tensor
S	total electron spin quantum number of an atom
t	time
T	absolute temperature
T_1	spin–lattice relaxation time
T_2	spin–spin relaxation time
x, y, z	Cartesian coordinates
ε	dielectric constant
ε_0	dielectric constant of free space
ϑ	angle between magnetic field and symmetry axis of the crystal filed
λ	spin–orbit coupling constant
μ	magnetic moment
μ_N	nuclear magneton
μ_B	Bohr magneton
ν	frequency
χ	complex paramagnetic susceptibility
χ'	real part of χ
χ''	imaginary part of χ
χ_0	static Curie susceptibility

BIBLIOGRAPHY

Literature for introductory and additional studies in the field is cited below. In order to concentrate on defects in solids, a large part of spin resonance work has been totally omitted in this book. Some books concerning these special topics are listed together with the more general literature.

General literature on magnetic resonance

A. Abragam, *The principles of nuclear magnetism* (Oxford Univ. Press, London, 1961).

R. S. Alger, *Electron paramagnetic resonance* (Interscience, London, 1968).

S. A. Al'tshuler and B. M. Kozyrev, *Electron paramagnetic resonance* (Academic Press, New York, 1964).

H. M. Assenheim, *Introduction to electron spin resonance* (Hilger and Watts, 1966).

J. W. Orton, *Electron paramagnetic resonance* (Iliffe Books, London, 1968).

G. E. Pake, Nuclear magnetic resonance, *Solid State Phys.* **2** (1956).

G. E. Pake, *Paramagnetic resonance* (Benjamin, New York, 1962).

C. P. Poole, *Electron spin resonance* (Wiley, New York, 1967).

C. P. Slichter, *Principles of magnetic resonance* (Harper and Row, New York, 1963).

Literature on special aspects

A. Abragam and B. Bleaney, *Electron paramagnetic resonance of transition ions* (Clarendon, Oxford, 1970).

P. B. Ayscough, *Electron spin resonance in chemistry* (Methuen, London, 1967).

L. Ebert and G. Seifert, *Kernresonanz in Festkörpern* (Akademische Verlagsgesellschaft, Leipzig, 1966).

D. J. E. Ingram, *Free radicals as studied by EPR* (Butterworths, London, 1958).

H. S. Jarret, Electron spin resonance spectroscopy in molecular solids, *Solid State Phys.* **14** (1963).

G. Lancaster, *Electron spin resonance in semiconductors* (Hilger and Watts, 1966).

W. Low and E. L. Offenbacher, Electron spin resonance of magnetic ions in complex oxides, *Solid State Phys.* **17** (1965).

T. H. Wilmshurst, *ESR spectrometers* (Hilger and Watts, 1967).

AUTHOR INDEX

SUBJECT INDEX

219